Development of Specifications for Biotechnology Pharmaceutical Products

•••••••••••••••••••••••

Developments in Biological Standardization

Vol. 91

This series Dev Biol Stand begins with Vol. 23 and is the continuation of both «progress in Immunobiological Standardization, Vols 1-5» and «symposia series in Immunobiological Standardization, Vols 1-22».

KARGER

Basel · Freiburg · Paris · London · New York ·
New Delhi · Bangkok · Singapore · Tokyo · Sydney

Development of Specifications for Biotechnology Pharmaceutical Products

Hyatt Regency, San Francisco CA, USA
May 2-3, 1996

Volume Editors

Fred Brown
Plum Island Disease Center, Greenport NY, USA

James Fernandez
Berlex Biosciences, Richmond CA, USA

Proceedings of a Symposium organized and sponsored by the
PDA West Coast Chapter and
International Association of Biological Standardization (IABS)

21 figures and 16 tables, 1997

KARGER

Basel · Freiburg · Paris · London · New York ·
New Delhi · Bangkok · Singapore · Tokyo · Sydney

Developments in Biological Standardization

Bibliographic Indices. This publication is listed in bibliographic services, including Current Contents® and Index Medicus.

Drug dosage. The authors and the publisher have exerted every effort to ensure that drug selection and dosage set forth in this text are in accord with current recommendations and practice at the time of publication. However, in view of ongoing research, changes in government regulations, and the constant flow of information relating to drug therapy and drug reactions, the reader is urged to check the package insert for each drug for any change in indications and dosage and for added warnings and precautions. This is particularly important when the recommended agent is a new and/or infrequently employed drug.

Printed in Switzerland by Médecine et Hygiène, Genève
ISBN 3-8055-6569-0

Organization of the Symposium

Session I:	Global Perspective on Specification for Biotechnology Products Moderator: *Robert Baffi, Ph.D.*
Session II:	Evolution of Specifications Through Product Development Moderator: *Vincent Anicetti, M.S.*
Workshop I:	Environmental Control
Workshop II:	Specifications: What is Necessary to Start Clinical Trials?
Session III:	Specifications Within the Manufacturing Operation Moderator: *Florian Horaud, IABS*
Workshop III:	In-Process Testing/Limits
Workshop IV:	Reference Standards/Materials
Session IV:	Role of the Pharmacopoeia and Validation in Specification Development Moderator: *Jean Huxsoll, M.S.*

Scientific Committee

Robert Baffi
Genentech, Inc.

Kristen Bacigalupi
Chiron Corporation

Beverly G. McCalla
Chiron Corporation

Steven Carlson
Matrix Pharm., Inc.

James Fernandez
Berlex Biosciences

Robert Garnick
Genentech, Inc.

Elwyn Griffiths
World Health Organization

Florian Horaud
Institut Pasteur

Jean Huxsoll
Matrix Pharm., Inc.

Paul Kleinman
Biopharmaceutical Writers Service

Anthony Lubiniecki
SmithKline Beecham Pharm.

James C. Lyda
PDA

Jennifer Quentin
Microbiological Associates

Kenneth Seamon
FDA

Barabara Sheng
Genentech, Inc.

Randall Tedder
Mensco, Inc.

Introduction

On May 2 and 3, 1996 the West Coast Chapter of the PDA (WCCPDA) and International Association of Biological Standardization (IABS) sponsored a symposium titled Development of Specifications for Biotechnology Pharmaceutical Products. The meeting was attended by over 400 scientists and participants who listened to presentations on development of specifications from the IND stage through the marketed product stage from both industry and regulatory perspectives. Speakers from throughout the world participated in the presentations and many of the speakers were also a part of the International Conference on Harmonization meetings that preceded the symposium and were hosted by the WCCPDA.

Presentations on specifications from an industry perspective focused on the degree of specifications that are appropriate during various stages of drug development. Also included were talks on the development of specifications for raw materials, in-process materials and final product as it is gradually developed from rough formulation to a final, highly purified, biotechnology product. Definitions of which specifications are relevant; i.e. identity, purity, or activity at the various stages of drug development were presented and discussed in workshops that concluded each of the two day's meetings. Validation and its effects on the setting of specifications was discussed with the note that setting of specifications is always directly effected by the quality of the validation. Standards and reference materials used in determining specifications were discussed noting that the quality of the reference material used in defining standards is critical to the proper selection of a standard proposed for wider use in global settings.

The meetings concluded with a presentation of the workshop highlights where all of the symposium attendees had the opportunity to comment on the presentations and provide input to the ICH process of Specification, Tests and Procedures for Biotechnological/Biological Products (ICH Q6B.)

This volume of Proceedings includes the written presentations from the speakers that participated in the Symposium. The Proceedings are preceded by the Symposium Agenda with speaker names and titles. Also included are summaries of the workshops held during the meeting. The editors are very grateful to the authors for their timely contributions. We regret that some of the speakers did not submit manuscripts for inclusion in these Proceedings.

Many thanks are due to the organizers of the symposium. From the PDA, note should be given to Ed Fry, Jim Lyda, and the staff at the PDA offices. From the West Coast Chapter of the PDA acknowledgment should be given to Vince Anicetti for his skilful chairmanship of the group of volunteers that coordinated the innumerable tasks that were required to present a smooth meeting. From the IABS, the enthusiastic participation of Florian Horaud and Fred Brown should be appreciated by those who participated in the meeting and are reading this copy of the proceedings.

Jim Fernandez,
Berlex Biosciences
Richmond, California, USA

Contents

Brown F, Fernandez J (eds): Development of Specifications for Biotechnology Pharmaceutical Products. Dev Biol Stand. Basel, Karger, 1997, vol 91, p 1

SESSION I

Global Perspective on Specifications for Biotechnology Products

Moderator: *Robert Baffi, Ph.D.*

Brown F, Fernandez J (eds): Development of Specifications for Biotechnology Pharmaceutical Products. Dev Biol Stand. Basel, Karger, 1997, vol 91, pp 3-13

FDA Perspective on Specifications for Biotechnology Products – from IND to PLA

G. Murano

Center for Biologics Evaluation and Research, Food and Drug Administration, Rockville, MD, USA

Key words: Specifications, characterization, validation, identity, purity, potency.

Abstract: Quality standards are obligatory throughout development, approval and post-marketing phases of biotechnology-derived products, thus assuring product identity, purity, and potency/strength. The process of developing and setting specifications should be based on sound science and should represent a logical progression of actions based on the use of experiential data spanning manufacturing process validation, consistency in production, and characterization of relevant product properties/attributes, by multiple analytical means. This interactive process occurs in phases, varying in rigour. It is best described as encompassing a framework which starts with the implementation of realistic/practical operational quality limits, progressing to the establishment/adoption of more stringent specifications. The historical database is generated from preclinical, toxicology and early clinical lots. This supports the clinical development programme which, as it progresses, allows for further assay method validation/refinement, adoption/addition due to relevant or newly recognized product attributes or rejection due to irrelevance. In the next phase, (licensing/approval) specifications are set through extended experience and validation of both the preparative and analytical processes, to include availability of suitable reference standards and extensive product characterization throughout its proposed dating period. Subsequent to product approval, the incremental database of test results serves as a natural continuum for further evolving/refining specifications. While there is considerable latitude in the kinds of testing modalities finally adopted to establish product quality on a routine basis, for both drugs and drug products, it is important that the selection takes into consideration relevant (significant) product characteristics that appropriately reflect on identity, purity and potency.

Introduction

In the last decade advances in biotechnology have provided unique opportunities to design and construct molecules with desirable qualities to perform specific biological functions. Several of these «designer» molecules (many endowed with multiple pre-selected functions and defined specificities) have been developed and produced in large enough quantities to meet considerable medical demands spanning diagnosis, prophylaxis and therapy. Applications have spanned a rich and safe experience to include the management of metabolic and cardiovascular disorders, immune deficiencies, malignancies, anaemia, growth disorders, transplantation and auto-immune diseases. The inordinate complexity of the technology has rendered obligatory that uniform (harmonized) quality standards be adopted and applied throughout development, approval and post-marketing phases of biotechnology-derived products, thus assuring product identity, purity, and potency/strength.

Operationally, these quality standards are achieved and maintained by the development and adoption of a robust control strategy. Specifications – a legal commitment with all regulatory authorities – constitute a strategic component of a complex algorithm. To this extent, the process of developing and setting these boundaries of product performance throughout the dating period of the product should be based on sound science. This process should represent a logical progression of action driven by experiential data spanning manufacturing process validation, raw materials and in process controls, consistency in production, (lot-release testing), characterization of relevant product attributes by multiple analytical modalities and stability evaluation. In general, this interactive process occurs in phases which vary in rigour and varies depending on the nature of, or the step in, the manufacture of a product. It encompasses the development and adoption of realistic operational quality limits and progresses to implementation of more stringent specifications. The historical data base is generated from preclinical toxicology studies and early clinical experience. It serves in support of the pivotal clinical development programme which, as it progresses, allows for further assay validation/refinement, adoption/addition due to relevant or newly recognized product attributes or rejection due to redundancy or outright irrelevance. In the next phase (licensing/approval), specifications are set through extended experience and validation of both the preparative and analytic processes to include availability of and comparison to suitable reference standards and extensive product characterization throughout its proposed dating period. Subsequent to product approval, the incremental database of test results provides an opportunity for performing meaningful statistical assessments and serves as a natural continuum for further evolving/refining (generally tightening) specifications. Throughout this highly interactive process, the guiding principle driving the decision to adopt specifications at various rationally pre-selected points of manufacture (i.e. in process, purified drug substance, drug product) or for certain elements to rely on process validation, is the assurance that product consistency and quality are unbreached.

While there is considerable latitude in the kinds of testing modalities and point of testing finally adopted to establish product quality on a routine basis, for both drug substance and drug product, it is important that the selection takes into consideration relevant (significant) product characteristics that appropriately reflect identity, purity, and potency.

It is paramount to be highly sensitive to the seldom breachable linkage which exists between specifications and production process, stability profile, analytical methodology, materials used in preclinical and pivotal clinical trials and statistical validity of data analysis (Fig. 1).

Product Characterization

A rational choice of appropriate specifications is founded on thorough analytical and biological characterization of the molecular entity under consideration.

	• PRODUCTION PROCESS
	• STABILITY PROFILE
SPECIFICATIONS ⇐⇒	• ANALYTICAL METHODOLOGY
	• QUALITIES OF MATERIALS USED IN PRECLINICAL AND CLINICAL TRIALS
	• STATISTICAL VALIDITY OF DATA ANALYSIS

Fig. 1: Parameters constituting the obligatory linkage to specifications.

Extensive physicochemical characterization need be determined once only and, together with in-process defined parameters, most specifications should be based on them. A characterization programme will generally include a definitive determination of the composition, physical properties and primary and higher order structure (amino acid analysis, carbohydrate analysis, amino acid sequence and, where feasible, disulphide bond structure and carbohydrate sequence); a thorough assessment of its molecular weight, extinction coefficient, iso-electric point, higher order structure by UV/visible spectrum, CD, IR, NMR, crystal structure; sub-unit structure by reducing vs. non-reducing SDS-PAGE. Where useful, assessment of unique, identity-indicating patterns obtained with various forms of high resolution analytical techniques (peptide map, chromatographic profiles, IEF, SDS-PAGE, CE, MS) may be appropriate.

Specifications are chosen to confirm, rather than establish, molecular structure and should focus on those molecular characteristics found to be useful in ensuring the safety and clinical utility of the product. Thus, the set of tests chosen will vary from product to product and stage of manufacture and should be guided by preclinical and clinical safety and effectiveness data as much as possible (Fig. 2).

PRODUCT CHARACTERIZATION

- composition/structure
  - a. a. analysis/sequence
  - CHO composition/sequence
  - S-S
- physical properties
  - MW, Ext. Coeff. IEP.
- higher order structure by UV Spec., CD, IR, NMR, crystal
- subunit structure by SH/SS-SDS PAGE
- identity-indicating patterns
  - peptide map
  - chromatography profiles
  - IEF, SDS-PAGE, CE, MS

Fig. 2: Summary of a typical product characterization profile.

Physicochemical results alone are not sufficient to characterize a protein satisfactorily. Relevant biological characterization test(s) and assay(s) which reflect the intended mechanism of action should be performed to assess the biological characteristics of the product. Selection should be based on the ability of the modalities to provide maximal information predictive of clinical use. For example, for monoclonal antibodies, specificity should be evaluated by a combination of immunochemical and/or functional assays. Direct binding assays to purified antigens and defined regions of antigens, and affinity, avidity and immunoreactivity determinations should be performed, as feasible. Whenever possible, the target molecule bearing the reactive epitope should be biochemically defined and the antigenic epitope itself determined.

Process Validation and Controls

Process validation and controls constitute key elements of the overall strategy in assuring product quality. It provides documented evidence that the process is effective and reproducible and yields a product that consistently meets pre-determined quality specifications and attributes. Process validation encompasses the development, implementation, and ultimate performance of a specific process or process step. It includes the establishment of initial goals for specifications and quality attributes. It progresses through development in the laboratory and pilot plant where the process is refined, critical parameters determined, equipment selected, installed, qualified, acceptable operating limits and performance established, and demonstrated to perform at a commercial scale. As part of process validation, critical process parameters are established and evaluated to ensure the robustness and consistency of the process.

Process controls encompass a combination of detailed manufacturing instructions, analysis of selected samples at points in the process where important information is obtainable for the establishment of meaningful action limits. Action limits are based on design criteria and validation studies. Excursions above action limits do not necessarily indicate that product quality has been compromised, but rather a need to evaluate the significance of that excursion in the light of available data, to investigate its cause and initiate action to restore and maintain desired limits of control.

Identity

The identity test procedure may be defined as a method or group of test methods, based on previously derived analytical results, which uniquely identifies(s) the molecular structure of the active drug component. The identity test is not a complete molecular characterization of the drug substance, but should be able to detect subtle alterations in its molecular structure. Identity is generally a qualitative measure of the physical and chemical structure of the drug substance which favours specificity over sensitivity.

Identity test(s) generally will be performed at appropriate points in the manufacturing process and the tests may differ depending on the stage of manufacturing. Complete identity testing should be performed on the purified drug substance of all the lots which: (i) are used clinically; (ii) are used for the production of the in-house reference standard; (iii) are used to support consistency, and (iv) are prepared for release after licensure. For monoclonal antibodies, identity tests must verify: (i) the Ig class/subclass (H+L); (ii) the primary molecular structure; (iii) subunit composition/molecular integrity; (iv) conformational integrity; (v) specificity for the target antigen, and (vi) consistency of post-translational modifications. It is not necessary to analyse quantitatively the mono/oligosaccharide content of monoclonal antibodies. However, since glycosylation and other post-translational modifications may have an impact on the activity and pharmacokinetics of monoclonal antibodies, these modifications should be controlled. For immunoconjugates, the average ratio of coupled material to antibody should be determined, and the heterogeneity of conjugated species should be analysed and specifications set to control for consistency.

Identity tests should be specific for each drug product, and identify it as the product designated on the final product label. Tests should distinguish the drug product from other drug products manufactured at the same facility, to prevent mixups and/or mislabelling (Fig.3).

Purity and Impurity

Purity should be determined by at least two or more methods which operate by two different principles of separation. At least one method should be quantitative. The choice and optimization of methods should focus on separation of the active

IDENTITY

- test(s) uniquely identify molecular structure
  * generally qualitative measure
  * favours specificity over sensitivity
  * not complete molecular characterization
- for drug substance
  * performed at appropriate points in manufacture
  * may differ depending on stage
  * complete id on all lots:
    * clinical/reference standard/consistency/marketing
- for drug product
  * specific/adequately id as designated on label
  * distinguish from other products
    e.g. produced in same facility

Fig. 3: Summary of principles and applications of Identity Testing.

drug substance from impurities. Since the analytical value obtained for purity of a given preparation depends upon the analytical method used, relevant lower limits should be established for each method. The value set for the purity limit should take into account analytical data from clinical trial batches, manufacturing capabilities, method variability, and stability studies. The analytical method descriptions should include a clear explanation of the calculations used for purity determinations (Fig. 4).

Sensitive methods should be used to determine the level of impurities. As in the case of purity determination, actual values depend on the method(s). Impurities may be either of known structure, partially characterized, or unidentified. The identity and potential toxicity of all detectable impurities should investigated. Unless various impurities are isolated and individual absorption coefficients determined, the analytical values obtained for their levels in the drug substance are necessarily estimates. Impurities that are selected for inclusion in the tests and specifications are «specified impurities». The rationale for their inclusion or exclusion and a comparison of the impurity profiles of clinical batches with those proposed for marketing should be presented. There are many examples of commonly encountered product-related impurities.

a) Truncated forms. Cellular peptidases may catalyse the removal of amino acids or catalyse internal cleavages. For small proteins, these may be detected by HPLC or SDS-PAGE. For very large proteins, HPLC or SDS-PAGE may offer insufficient resolution. Peptide mapping is the method of preference.

b) Deamidated, oxidized, improperly cross-linked, isomerized forms may require considerable effort in isolation and characterization to identify the type of chemical modification(s) and amino acid residue(s) involved. Chromatographic and/or electrophoretic methods (e.g., reHPLC, capillary electrophoresis) may be used to isolate and identify such variants.

PURITY

- determined by two methods defined by two principales of separation
- one should be quantitative
- specific limits set for each methos based on
  - analytic data from clinical trial lots
  - manufacturing capabilities
  - method variability
  - stability experience
- explanation of calculations

Fig. 4: Summary of purity-establishing profiles.

c) The category of aggregates includes dimers and higher multiples of the active moiety. These are generally resolved from the active moiety and quantified by size exclusion chromatography (e.g., SE-HPLC).

d) Degradation products identified from accelerated or real-time stability studies as being generated in significant amounts should be included in the specifications profile.

Upper limits should be established for the molecular variants described. There are many examples of process-related impurities.

a) Reagents that have not been otherwise demonstrated by process validation studies to be removed from the drug substance during the manufacturing process should be included in the specifications profile. Of particular concern are antibiotics, antibodies, and enzymes which may produce undesirable responses in humans. For immunoconjugates, the amount of free antibody and free conjugate component in the final product should be determined and limits set for each.

b) Expression organism components and media components. A sensitive immunoassay capable of detecting a wide spectrum of protein impurities is generally used. The polyclonal antibody used in the test is generated from a crude preparation of a mock production organism, i.e. a production cell minus the product-coding gene.

c) The manufacturing process should be designed to result in a product that is free of infectious agents (viruses, bacteria, fungi, mycoplasma). The ability of the manufacturing process to remove/inactivate replication-competent retroviruses should be validated.

d) Other contaminants that are detectable in the drug substance and that have not been otherwise demonstrated by process validation studies to be removed from the drug substance during the manufacturing process should be included in the tests and specifications. These include solvents, anti-foam agents, chemical additives, leachates, and heavy metal ions.

Where possible, specific impurities reference standards should be used for quantifying individual impurities. Specific upper limits which take into account analytical data from clinical trial batches, manufacturing capabilities, method variability, and stability studies should be established for the specified impurities. The limit should also be established for specific, identified, product-related impurities. The limit established for specific, identified, product-related impurities may depend on whether these are biologically active. In addition, a specific upper limit should be established for total impurities, which includes unidentified species. Limits should be set no higher than can be justified by safety data and no lower than historically achievable by the manufacturing process and analytical capability.

For Lot and Batch release, the purity should be determined by two or more methods defined by different principles of separation. Routine lot release should include testing for impurities which, from stability studies, are known to increase substantially with time (e.g., aggregates by SE-HPLC, chemical variants by RP-HPLC) (Fig. 5).

IMPURITIES

- Product related
- Process related
- Efforts to identify and investigate toxicity of all detectable impurities
- Rationale for including/excluding specifications
- Standards (used for quantification)
  - specific limits for individual and total impurities
    - set based on whether species is biologically active
    - set no higher than justified by safety
    - set no lower than historically achievable
  - in absence of safety concerns, limits set based on manufacturing experience
    - with sensible latitude
- Lot release
  - Purity determined by two methods defined by different principles of separation
  - Impurity determined based on experience from stability studies (e.g. degradants)

Fig. 5: Summary of Impurity-establishing profiles.

Biological Activity

Determining the biological activity of well-characterized biotechnology-derived therapeutic proteins and monoclonal antibodies assures that the product will have the expected clinical effect. Its specific activity reflects the purity of the lot. Furthermore, the bio-assay result functions as a surrogate for the integrity of the higher order structure of a protein.

The methodologies used to measure biological activity can be classified into three categories, animal assays, cell line-based assays, and biochemical assays. While it is desirable and sometimes necessary to use a biomimetic assay (an assay that measures the required biological effect), in fact, this practice is often difficult.

Animal assays are time consuming, costly and intrinsically imprecise, but yield accurate results when a large number of animals is used. When an animal bio-assay is used for lot/batch release, one must accept a wide potency range. A bio-identity test is an assay using fewer animals than normally required in a full bio-assay and is not capable of determining the biological activity of a product with sufficient statistical power; it is useful as a limit test confirming that the product is biologically active and has a specific activity that is not below an established value based on assay variability. A bio-identity limit test alone is not sufficient for product characterization or for the quality control of the reference standard. For products with a long history of clinical use, a bio-identity limit test may substitute a true bio-assay for lot/batch release. A widely used method for the determination of biological activity of a protein involves using stable cell lines. In general, the cell line-based assays in which the modulation of signal transduction occurs in response to the target molecule (which in turn results in a meaningful biological response cell line-based bioassay) is preferred over assays based solely on receptor binding (cell line-based binding assay). The latter are acceptable only when scientifically justified. The stability of the cell-lines used for these assays is critical. Once validated, cell line-based assays are comparatively more precise, less time consuming, and less costly than animal assays. Potency specifications with reasonable ranges can be established based on assay variability, and data of preclinical and clinical lots/batches.

Biochemical assays do not involve the use of a living biological system but are based on either the binding of the product to other substances, or a chemical action (function) of the product itself. Examples of the binding-based biochemical assays are immuno-assays, isolated receptors-binding assays, and substrate-binding assays. In general, receptor-binding assays and immuno-assays (especially using polyclonal antibodies against denatured proteins) are not acceptable for therapeutic proteins unless exhaustively characterized. On the other hand, immuno-assays are quite adequate for the determination of the biological activity of monoclonal antibodies. Substrate-binding or functional-based biochemical assays are also acceptable for most enzyme drugs. Once validated, these biochemical assays are easy to perform, are precise, and indicate stability. Consequently, specifications with reasonably narrow ranges can be established. It should be noted that when the clinical effect of an enzyme or monoclonal antibody involves tissue targeting or is based on signal transduction of a specific type of cells, a cell culture-based binding assay or bioassay should be used for potency determination.

Potency assays for monoclonal antibodies should reflect, to the extent possible, the intended mechanism of action of the mAb in the human recipient. This should include a combination of analyses that evaluate binding activity, biological activity, and biochemical activity, as appropriate. Antibody binding should be characterized for affinity, avidity, and immunoreactivity, to the extent possible. Quantitative methods that are sensitive to these parameters, either individually or in combination, may be used. Compound antibodies or conjugates possessing more than one function must be shown to possess both activities on the same molecule. Specific activity should be expressed as unit activity per mg of product.

Finally, whenever the degree of the assay variability exceeds the lot-to-lot variation in true specific bio-activity, it is preferable to quantify the product for placing in vials and patient-dosing on the basis of protein mass which can be measured accurately (Fig. 6).

BIOLOGICAL ACTIVITY

- Bio-assay
  - * assures expected pharmalogical response
  - * S.A. reflects particular lot
  - * surrogate for 2nd order structure
- Methodologies: preference for biomimetic/clinically relevant assay, ie. measures required biologic effect
  - * animal assay
  - * cell line-based assay
  - * biochemical assay

If assay variability exceeds lot-to-lot variation, quantify based on mass.

- Standards
  S.A. determined by one or multiple assays
- Lot release
  - * drug substance/Product
    - Assay selection based on principles involved
    - Specifications set based on
      - assay variability
      - data on early lots
      - may be modifed with incremental experience (statistically valid)

Fig. 6: Summary of Biological Activity-establishing profiles.

QUANTITY

A carefully validated specification for quantity (also referred to as potency, mass, or strength) is essential to guide formulation and ensure appropriate dosing. Analytical determination of protein mass is far more precise than most biological assays and therefore should be the basis for this specification. A variety of procedures (e.g., extinction coefficient, dye-binding assay, quantitative amino acid analysis, etc.) may be used, but should be validated against a reference method (nitrogen determination, quantitative amino-acid analysis, gravimetric analysis). If a procedure other than a reference method is chosen, a set of standards prepared from reference material should be validated.

In general for Lot and Batch Release, quantification of the drug substance is uncomplicated by the presence of carrier protein or other stabilizers that may interfere with the protein assay chosen. When carrier protein or other substances (e.g. reducing agents, chelating compounds or detergents) must be added to the final formulation, a method should be chosen that can tolerate the additional product components, and this method should be appropriately validated. For example, when a final product contains carrier protein, an assay based on specific immunoreactivity of the active material might be appropriate.

STATISTICAL CONCEPTS

Appropriate statistical analyses should be applied to any quantitative results. The intent of these analyses is to determine the degree of confidence one can place on such determinations. A description of the statistical methods used should include the following elements:

(i) A definition of the variables involved. Examples would include: amount of drug substance or product in milligrams, location (e.g. elution time) for a drug substance or impurity, and area under chromatographic peak for a drug substance or impurity.

(ii) A statement of what constitutes a valid experimental procedure or assay.

(iii) A detailed description of the statistical method. This should include a statement of the assumptions of the procedure such as normality, linearity of response, transformations, etc. For example, in determining purity, a peak must be present at a specified location. For substances exhibiting multiple peaks, a peak-by-peak profile should match a reference standard. In potency determinations, a bio-assay method will provide (log) relative potency (rp) and the standard error of the (log) rp.

(iv) A report of the statistical analyses and conclusions. This should include the results of the statistical analysis and not just a statement of «pass or fail». It is useful to evaluate the applicability of the statistical procedure. All statistical procedures have associated assumptions. The data from the specification procedures should be used to verify assumptions. If the assumptions are not satisfied, a better statistical model should be developed.

Periodically, the elements described above should be updated based on the history of the specification process. For example, in an assay performed for lot release, the standard deviation of (log) rp must be estimated. At the time of drug registration, an estimate of the standard deviation is provided by the assays used in the product licensing approval process. Re-evaluating these at regular intervals using all additional available data can provide a better estimate of the standard deviation of the assay.

Conclusion

This review summarised the impressive array of products deployed in various forms of medical management – a consequence of the burgeoning activities emanating from biotechnology in the past dozen years or so. To the credit of the biotechnology industry and the medical and patient communities is the equally impressive record of safety and efficacy associated with the use of these products. One is left with the impression that, operationally, the merits of this success story are founded not only on the clever and innovative technical developments so studiously nurtured by all, but also on the profound vigilance devoted to ensuring the highest quality. Allthrough the title of this review implies an FDA perspective on specifications, it is instead a documentation of the industrial experience which, while at times varying on the details of the approaches in developing and setting specifications, has always been concordant with the principles and usefulness of this all-important activity.

Acknowledgements

The perspective summarized in this presentation represents the collective effort of the FDA working group on Development of Specifications for Biotechnology Products. I was honoured by the privilege of conveying this information in this forum. I am grateful to all members of the working group: N. Goldman, M. Moos, R. Nordan, W. Egan, K. Poffenberger, S. Stibitz, P. Lachenbruch, Y.Y. Chiu, S. Moore, D. Wu, J. Brown.

Dr. G. Murano, CBER/FDA, 1401 Rockville Pike, Rockville, MD 20852, USA.

Brown F, Fernandez J (eds): Development of Specifications for Biotechnology Pharmaceutical Products. Dev Biol Stand. Basel, Karger, 1997, vol 91, pp 15-23

Global Perspective on Specifications for Biotechnology Products – Perspective from Japan

T. Hayakawa

Division of Biological Chemistry and Biologicals,
National Institute of Health Sciences, Tokyo, Japan

Key words: Japanese perspective, biotechnology products, specifications, regulatory evaluation and control.

Abstract: Current trends in Japan concerning regulatory evaluation and control of biotechnology pharmaceutical products, especially the development of their specifications, is described briefly.

INTRODUCTION

This paper describes briefly the current trends in Japan concerning regulatory evaluation and control of biotechnology pharmaceutical products, especially the development of their specifications.

REGULATORY CONCERNS ABOUT BIOTECHNOLOGY PHARMACEUTICAL PRODUCTS

At present, the official regulatory position of the Japanese government concerning evaluation and control of drugs produced by biotechnology can be found in two notifications, PAB (Pharmaceutical Affairs Bureau of the MHW) Notification No. 243 originally issued in March 1984 which will be revised in the near future and PAB Notification No. 10 issued in June 1988. The former notification covers recombinant DNA (rDNA) peptide or protein products derived from micro-organisms; the latter notification covers rDNA products derived from mammalian cells, monoclonal antibodies, and any other protein drugs produced by certain cell culture systems.

They are expected to enable both regulatory scientists and manufacturers to arrive at a consensus, and to provide a sound scientific basis for certain consistent regulatory policies for the control of quality, safety, and efficacy of medical agents produced by biotechnology. Additionally, these Notifications emphasize that because there are various types, characteristics and clinical uses for biotechnology drugs, and because scientific advances and new experiences in this field are being gained on an almost daily basis, the items contained in these Notifications should not be regarded as being all-inclusive or definitive. It is therefore considered to be desirable that a flexible case-by-case approach be adopted when testing and evaluating individual biotechnology drugs, and this should be based on scientific rationale available at the time.

Recently, international harmonization of the technical requirements of the quality of biotechnology products has been sought by six parties from the EU, USA and Japan, as well as the Canadian government through the ICH process. We will incorporate the ICH results, such as harmonized guidelines, into our regulatory requirements. Thus, the overall quality control of biotechnology drugs will also be influenced by the ICH results.

General Considerations for Evaluating Biotechnology Products

When evaluating the quality relative to safety and efficacy of biotechnology products, certain aspects should be considered:

(i) It is essential that the manufacturer should provide a detailed description of the manufacturing process in order to verify the desired product. Validation of the ability of an adopted process to yield the product with certain specifications and to establish the consistency of that process are essential elements in providing the basis for an acceptable protein drug from biotechnology;

(ii) Extensive identification and characterization of the products by a combination of detailed physicochemical, immunochemical and biological tests are of major importance because they are the products of artificial manipulation.

(iii) Special concern should be paid to contaminants existing in the final product that may be hazardous to human health. Product-related substances will also be a subject of concern.

(iv) The established quality of a biotechnology drug during its developmental stage with respect to safety and efficacy must be controlled on a lot-by-lot basis after its approval as a new drug. Therefore, it is necessary to establish the appropriate specifications and testing methods that can most accurately reflect the special nature of the product.

The specifications should also be established by taking into account various factors such as specific features of the manufacturing process, results of the process evaluation studies on the clearance of possible contaminants, the results of stability studies, and preclinical safety and clinical studies.

The aforementioned quality control-related factors will be described in more detail.

Validation of Manufacturing Processes and Establishment of their Consistency

The production of drugs by biotechnology is fundamentally based on the artificial manipulation of the gene or a living organism, or both. There may still be unreliability within the manufacturing process. In addition, because biologically active protein products are usually unstable, they may be modified physically or chemically during separation, additional processing of the primary product, the purification process, formulation into dosage form, or storage. The possible production of a subpopulation of molecules with abnormal primary, secondary, or tertiary structure may result in adverse reactions, as well as undesirable effects on the efficacy of a desired drug. Detailed data on the manufacturing process from the establishment of initial cell substrate to the purification of the product should be provided, to verify drug production directed by a specific gene coding for the desired product in recombinant host cells or established cell lines. These include: verification of gene manipulation, if any; derivation, passage history and characteristics of the cell substrate; data on the results of tests for contaminating micro-organisms; cell bank system employed; stability of the cells during storage and times of drug production; culture conditions; isolation and purification processes of the product; stability of the product. Data on the impact of the formulation on the integrity of the drug substance should also be described.

For validation of the manufacturing process adopted, and for assurance of consistency of the manufacturing process, the ICH documents on cell substrate, genetic stability and viral safety shall be consulted as the primary important guidelines where possible and appropriate. However, in some cases, more detailed or additional information may be found in the national/regional guidelines/guidances and/or related documents.

Detailed information on the manufacturing process is also essential to define the nature and range of testing for the final product. The approach used by manufacturers in their overall strategy for validating their specific product and its manufacturing process should be explained and justified.

Analysis of Structural Integrity and Characterization of the Desired Product

The purified biotechnology product should be the subject of extensive identification and characterization. Because structural variants can be produced during biosynthesis or purification, there should be experimental confirmation that the protein product was actually preserved in its intended structure during the biosynthesis and purification processes. Whenever possible, the biotechnology product should be compared with the product extracted from natural sources.

Evidence for the structural integrity and characterization of the desired product comes from the results of structural analysis, as well as a wide variety of physicochemical, immunochemical, and biological tests.

The establishment of a relevant reference standard from the fully characterized lots of a given biotechnology product will be essential in assessing and controlling subsequent lots.

Analysis of Structure and Composition

Confirmation of the primary, secondary, and tertiary structures of a product by means of appropriate methods in protein chemistry, such as amino acid composition analysis, terminal amino acid (sequence) analysis, determination of the numbers and positions of sulphydryl group(s) & disulphide bridge(s), analysis of peptide fragments from digestion of the product, amino acid sequencing, analysis of the carbohydrate moiety of the glycoprotein products, analysis of high-order structure, and other relevant chemical and physical tests, are of major importance because the protein product is generally a complex substance produced by artificial manipulation of living organisms.

For primary structural analysis of the protein moiety of the product, automatic amino acid analysis, automated Edman degradation, and peptide mapping with the combination of automatic amino acid analysis, automated Edman degradation and/or mass spectrometry would be the most useful analytical methods to be employed.

Examination of Physicochemical Properties

To examine the physicochemical properties of the protein product, the Japanese authority currently suggested examples of some physicochemical properties to be studied and typical methods to be employed. These include: spectrometric characteristics (e.g., UV); isoelectric point (e.g., IEF); molecular weight (e.g., gel filtration, SDS-PAGE under reducing and non-reducing conditions); electrophoretic patterns (e.g., PAGE, IEF, SDS-PAGE, capillary electrophoresis (zonal-, micellar-, gel-)); liquid chromatographic patterns (e.g., SE-, IE-, RP-, HI-HPLC); high-order structure (e.g., nuclear magnetic resonance (NMR), circular dichroism (CD), X-ray analysis)

Examination of Immunochemical Properties

A variety of polyclonal and monoclonal antibodies are available, or can easily be raised, for many proteins and for some specific segments in the polypeptide chain of a protein. If the immunochemical properties of the desired protein product are used in tests for identity, homogeneity, purity and assay, in the purification process of the product, or its pharmacokinetic studies, all related information should be made available.

For antibodies used for the above-mentioned purpose, their reactivities with their desired product and the methods of preparation should be described according to the use of the antibody.

Examination of Biological Properties

It is often difficult to assess the structure of a protein completely by only physicochemical means. It is particularly difficult to assess the conformational aspects of structure, which may be important both for biological activity and for immunogenicity in humans. Preservation of the correct structure relative to specific biological features of a protein during the production and purification processes of biotechnology products should be examined extensively. Bioassays may become powerful tools in the identification of complex proteins in their entireties and are most important for their potency estimates.

There are many biological substances that possess diverse actions on living systems. It is, therefore, desirable to characterize their biological features, and to elucidate their mechanism of action, as far as possible. At least, typical biological activities and biochemical properties which provide necessary information for identification and confirmation of the desired product should be described.

Special Considerations for Glycoproteins

Many medically useful glycoproteins have been produced by biotechnology and some have been approved for marketing in Japan. There is no doubt about the importance of proper characterization of glycoproteins relative to their structural and biological features.

One of the points to consider in testing or evaluating the glycoproteins produced by biotechnology as medical agents is that oligosaccharides attached to a protein may up-regulate, down-regulate, or turn off the biological activity of the glycoprotein in question. They may also affect the metabolic fate, stability or solubility of the proteins.

Another point to consider is that species-, cell-, and tissue-specific N-glycosylation is now firmly established, and the unique glycosylation capacity of the cell has been referred to as defining its glycotype. Expression of glycoproteins of one species in another, by genetic engineering, may lead to the isolation of variants of that glycoprotein. Even within the same cell type, the production of distinct sets of glycoforms, which differ in the fine structure of certain oligosaccharides and relative frequency of common oligosaccharides, may occur. Many factors, such as the method of establishment of the cell substrate, culture conditions used, or structural features of expressed protein products, can exert a certain influence on the nature and extent of glycosylation.

Proteins without the appropriate carbohydrate moieties may have altered pharmacokinetics and tissue distribution and may also cause other unknown and possibly adverse effects, including an antagonistic action against its natural form, immune complex formation through interaction with naturally occurring antibodies that recognize a specific carbohydrate structure, possible activation of monocytes, and interaction with various cellular receptors for its carbohydrate moiety. For products that are administered parenterally in large amounts or repeatedly over long periods, non-glycosylated proteins may reveal new antigenic sites and glycoproteins with unnatural carbohydrate structures may be antigenic.

In other words, biological, immunological, and/or physicochemical properties of glycoproteins may be controlled by the nature of glycosylation (structural feature of oligosaccharide chains), occupancy of glycosylation sequence and glycoforms at each glycosylation site. The site-specific functional role of the carbohydrate moiety may change in response to the derivatizing of its polypeptide backbone structure.

Analysis of Carbohydrate Moiety of Glycoproteins

Although the characterization of glycoproteins is by no means an easy task, recent progress in this field has enabled major parts of the carbohydrate structure to be analysed.

For the characterization of the structural features of the carbohydrate moiety of a glycoprotein product, it is essential to carry out, at first, carbohydrate composition analysis of the product. This can often provide certain useful information on the possible existence of N- or/and O-linked oligosaccharides as well as the types of oligosaccharides, for example, high mannose or/and a complex type. Analysis of the type of linkage of neuraminic acid as well as the nature of its capping (the ratio of N-acetylneuraminic acid content and N-glycolylneuraminic acid content) should also be performed. Glycosylation sites of the polypeptide chain, which are predictable in the case of N-glycosylation from the primary structure of the polypeptide backbone, should be determined by appropriate analytical procedures such as amino acid sequence analysis of isolated protease digests of the glycoprotein. The peptide fragments which contain carbohydrate moieties should then be the subject of carbohydrate composition analysis. This can provide information with respect to the type of oligosaccharide of a given glycopeptide. Preliminary structural examination of glycoforms of major oligosaccharides which may differ in the nature and number of branch, size and distribution can be performed by relatively simple analytical methods such as two- or three-dimensional oligosaccharide mapping. Where possible and appropriate, it is desirable to carry out further characterization of the carbohydrate structure, including glycosyl linkage composition, the degree of branching and sequence analysis. For some glycoprotein products, the structural feature and the extent of the heterogeneity of the oligosaccharide chains at each glycosylation site, as well as the differences between the N-linked carbohydrate chains among several attachment sites, have already been elucidated.

It should be emphasized again that the accumulation of data on the carbohydrate structures of glycoproteins results in a clearer elucidation of their role in the biological profiles of glycoproteins as therapeutic agents. As a result, it may become possible to select or produce more useful protein drugs having appropriate carbohydrate moieties. The extent of characterization of glycoproteins, as well as the control of their quality in relation to the effects of carbohydrate moieties on safety and efficacy, should be one of the more important issues to be discussed.

Analytical Methods of Oligosaccharide Structure

Some examples of carbohydrate analytical methods that may be useful for structural analysis of the oligosaccharide moiety of the glycoprotein products are as follows;

For monosaccharide analysis: chemical analysis, gas chromatography (GC); fluorescent-tag labelled monosaccharide-HPLC; high performance anion exchange chromatography (HPAEC); complex matrices with GC/MS or stand-alone tandem mass spectrometry.

For oligosaccharide mapping (two dimensional or three-dimensional): PA – oligosaccharide (anion-exchange, hydrophobic and hydrophilic interaction modes).

For oligosaccharide structure: electrophoresis (fluorophore-assisted-carbohydrate-electrophoresis: FACE, high-performance capillary electrophoresis: HPCE); combination of various methods (e.g., sequential exoglycosidase digestions, methylation analysis, controlled acetolysis, endo- and exoglycosidase treatment, methylation combined GC/MS analysis, LC/MS, FAB/MS, NMR); FAB/MS; ES/MS; NMR.

Possible Extraneous Contaminants and Product-related Impurities

In addition to extensive characterization of the biotechnology product itself, it is necessary to examine potentially hazardous contaminants that may be present in the final product. Possible extraneous contaminants and product-related impurities include: microbial contamination; pyrogens; cell constituents (e.g., cellular proteins, DNA, other constituents); culture medium constituents; substances from product-processing steps; substances from product purification steps (antibodies from affinity column, other reagents); product related impurities (e.g., mutants, erroneously cleaved products, disulphide isomers, aggregates: dimers & polymers, chemically modified forms: deamidated or oxidised forms, other degradation products).

Tests for the absence of contaminants, such as microbes and pyrogens, which are known to cause adverse reactions in humans, should be very rigorous.

By using reliable and sensitive tests, information on the types and quantities of other extraneous trace contaminants and product-related impurities in the final product should be provided. It may be necessary to set permissible limits for some contaminants in the final product, as a specification of lot-by-lot testing.

Process Validation for Elimination of Notable Impurities

For some impurities which may be present in the final product and would be subjects of quality and/or safety concerns, the effectiveness of the purification process with respect to the removal of impurities in question should be described. Process validation for elimination of notable impurities will be very important in deciding the nature and range of testing for the final product.

By eliminating or limiting the amount of recognized impurities, we will be able to avoid hazards specific to these agents. Fewer contaminants mean greater safety for the final product, a fact that is especially important for products administered repeatedly or in large doses.

Product-related Substances

Among the product-related substances, there are those that are known to retain similar biological activity to the native one, for example, the desamido and sulphoxide derivatives of hGH or desamido derivatives of insulin. They may be regarded as one of the active ingredients, but permissible limits should be defined.

Establishment of Specification and Testing Methods

The established quality of a biotechnology drug during its developmental stage will be controlled on a lot-by-lot basis after its approval as a new drug. Therefore, it is necessary to establish the appropriate specifications and testing methods that can most accurately reflect the special characteristics of the product.

When establishing specifications for routine batch control, variation in experimental estimates among defined numbers of batches and variation in experimental estimates from the same analytical procedure should be considered. Additionally an appropriate statistical treatment should be used.

The testing methods should be validated according to the ICH guideline «Validation of Analytical Methods», except in cases where there are specific issues for unique tests using analysing biotechnological products. The specifications should also be established by taking into account various factors such as, specific features of the manufacturing process, results of the process evaluation studies on the clearance of possible contaminants, the results of stability study, and preclinical safety and clinical studies.

Manufacturers should supply supportive data and explanation as well as overall details for setting specifications.

Selection of Batches

The selection of batches is also important. According to our guidelines, data from at least three batches produced under the conditions that reflect commercial-scale production should be the basis for setting specifications for both drug substances and drug products. At present, in Japan there is no rule or official position concerning specifications required for starting clinical trials, although a GMP guideline on medical agents intended for use in clinical trials is currently being prepared.

Examples of Items

The present Japanese guidelines indicate examples of the items which need to be considered for describing and testing after approval of the drug substance and drug product. The following show the case of drug substances: origin and nature; properties; identification; amino acid composition; peptide mapping; carbohydrate composition for glycoproteins; testing for purity; endotoxin test or pyrogen test; assay(s). Exemption from analysis of amino acid composition and peptide mapping may be obtained if the primary structural identity can be sufficiently confirmed by employing other test methods. Endotoxin or pyrogen tests should be performed when appropriate.

The case of drug products are as follows: origin and nature; properties; identification; pH; osmolarity; loss on drying or water determination; test for insoluble matter; weight variation test or actual volume deviation test; endotoxin test or pyrogen test; sterility; assay(s).

The items exemplified here in case of drug substances and drug products are not always required for every specific compound/product. Applicants should select those considered to be necessary for individual compounds/products from a scientific viewpoint.

Upon New Drug Approval in Japan

Upon new drug approval in Japan, a wide variety of scenarios with respect to the details of the types of manufacturing processes, testing items, methods as well as specifications can be left to the manufacturer's discretion if it is judged by the MHW to satisfy the purpose of ensuring the quality control of the entire lot. Thus, the specifications and testing methods of already approved biotechnology drugs, not only a specific individual drug but also identical or closely related biotechnology drugs, differ in the nature of the details and the relative weight of each testing item, as well as in the specifications.

HARMONIZATION

Since this year, the six parties from the EU, USA and Japan, as well as the Canadian government have started to address international harmonization of technical requirements for biotechnology product specifications. Japanese regulatory authorities will make an all-out effort to achieve such an objective through the ICH procedure. It is expected that the ICH's active efforts, based on scientific rationale, will lead to the development of a document that includes a guiding principle, recommended testing items and methods for suitable characterization of biotechnology protein products and methods for the development of their specification.

Dr. T. Hayakawa, Division of Biological Chemistry and Biologicals,
National Institute of Health Sciences, MHW, 1-18-1 Kamiyoga, Setagaya-Ku, Tokyo 158, Japan.

Brown F, Fernandez J (eds): Development of Specifications for Biotechnology Pharmaceutical Products. Dev Biol Stand. Basel, Karger, 1997, vol 91, pp 25-30

Development of Specifications for Biotechnology Products – Perspective from Europe

J.H. Trouvin

Agence du Médicament, Unité Biologie, Saint-Denis, France

Key words: Biotechnology products; specifications.

Introductory Remarks

Before presenting some views on how to determine specifications, I will make some introductory remarks about specifications.

* *Specifications, why and for what products:*

All products used and developed by the pharmaceutical industry have to undergo, at different times, quality controls and must thus meet specifications. It is important to remember that in the European community all medicinal products, be they chemical, biological or biotechnological, are released and distributed under the responsibility of the Marketing Authorisation Holder. Except for a few products (e.g. vaccines and blood products), the health authorities do not control batches of the finished product before their release on to the market. Thus, it is the Marketing Authorisation Holder's responsibility to release only batches that meet the specifications that are described and approved in the marketing authorisation file.

Specifications should be considered as objective criteria that can be imposed on the Marketing Authorisation Holder by the health authorities at any time. In particular, within the framework of post-marketing surveillance, the health authorities regularly inspect production sites and carry out quality controls on products available on the market.

* *Well characterised product or not?*

Specifications apply to all products, both simple and complex. A product does not have to be «well characterised» before specifications are fixed. On the contrary, specifications are all the more necessary when a product is poorly characterised or difficult to characterise. In other words, the notions of specification and characterisation are in no way mutually exclusive, especially when it comes to biotechnology products, for which the degree of characterisation must be taken into account when setting up specifications.

* *Specifications versus in-process controls*

Another point that has to be taken into consideration in setting specifications for biotechnology/biological products is that they are structurally more complex than chemical entities and, despite progress in analytical techniques, it is virtually impossible to check the overall molecular characteristics of such products routinely. Thus, end-product testing alone cannot guarantee the quality, safety and efficacy of each batch, particularly when the efficacy and/or safety depend on some specific features of the molecule.

This means that, while it is essential to respect release specifications, it is also necessary to take into account data on the production, purification and pharmaceutical formulation. On this matter, the European Guidelines on biotechnology products [1] include the following sentence that highlights the need for in process controls: *«Although comprehensive characterisation of the final product is essential, considerable emphasis must also be placed on "in-process" controls, a concept which has been highly effective in the quality control of vaccines prepared by conventional methods.»*

Thus, for a biotechnology product, it is important not only to meet the specification limits at the final product stage but also to meet all other in-process limits to assure that the overall quality and consistency of the production is achieved.

How to Set Specifications

From the manufacturer's point of view it is important to ensure that specifications are neither too numerous nor too demanding, so as to limit the risk of rejecting too many production batches. In contrast, the aim of the Health Authorities is to ensure that the controls undertaken before a batch is released effectively guarantee that the product has reproducible quality and efficacy from batch to batch. They will thus find a large number of stringent controls and specifications reassuring.

To answer the question of how to set specifications, I shall mainly discuss the specifications for two production stages: the active ingredient and the finished product. These specifications are set «at release» and, if necessary and justified, «at expiry».

Controls are generally divided into four categories:

- identity;
- purity;
- activity;
- other tests.

These four types of control ensure that the product has the expected purity and activity. Depending on the product, the «other tests» will include sterility, colour, pH, pyrogenicity, etc.

It is important to note that for biotechnology products, whereas one can agree on these four categories, the specifications and test items in each category must be envisaged case by case. Indeed, it is difficult to envisage a systematic approach and to impose a compulsory test list, which should apply whatever the product. There are several reasons for this imposition:

(i) Some methods can be used to verify more than one aspect at the same time. For example, HPLC analysis can provide information on both the identity of the product and its purity. Other methods (electrophoresis, biological assays) allow more than one physico-chemical or biological properties of the molecule to be checked at the same time. These overlaps must be exploited to the maximum to obtain as much information as possible with a reasonable «minimum» number of methods.

(ii) on the contrary, given the complexity of biotechnology products, a method cannot provide information about all characteristics of the molecule and thus, a whole range of tests will be necessary, together with a series of qualitative or quantitative limits, to obtain a relatively complete view of the product. It is the manufacturer's responsibility to justify the choice of test items, methods and limits.

To quote again the European guidelines on biotechnology products (1): *«A flexible approach to the control of these products should be adopted. ...Thus, the production and control of each product must be given careful individual consideration taking fully into account any special features.»*

It is thus impossible to draw a fixed and standardised list of tests to be performed, although some tests or methods can be suggested.

Many sources of information should be used in setting specifications. Without going into technical details on each, a few important points need to be considered:

Information from the analytical development

A great deal of information will be generated under «Analytical development»:

(i) Analytical methods are essential, whether physicochemical (HPLC, SDS-PAGE, MS, etc.) or biological (immunological tests, bioassays, etc.). Each method will have to be validated to determine its specificity, detection limit and accuracy. Without suitable analytical methods the product cannot be thoroughly characterised. But it is also the complexity of the molecule that will force the applicant to develop and use more and more sophisticated methods.

(ii) Thorough studies of the product and especially its molecular characterisation, in terms of primary, secondary, tertiary structure, glycosylation, carbamylation, etc. are needed to gain a good knowledge of the molecule. These analyses will focus on structures essential for activity in order to identify the most important characteristics such as disulphide bridges, glycosylation, N-terminal, amino acid etc. that have to be checked routinely as part of the test items.

(iii) The impurity profile, in relation to the production/purification process. A knowledge of the main impurities, whether «product-related» or «process-related», is necessary to define and justify the type and degree of impurities for routine controls. Some impurities do not have to be checked routinely once their elimination has been validated. One example is DNA testing, which is undertaken routinely on products derived from mammalian cells, while the elimination of bacterial DNA is simply validated and does not have to be checked routinely.

(iv) The stability profile enables all or some of the causes of instability or degradation to be anticipated, and identifies possible degradation products that may have to be checked routinely. Stability will have to be envisaged not only for active ingredient but also for the finished product, particularly to justify the proposed pharmaceutical formulation.

It is on the basis of this information that the applicant will be able to justify the different specifications at the release and expiry dates. It is important to stress that in Europe there are only «release specifications». The extension of limits at «expiry date» should be used in a limited number of circumstances and with clear justification and supporting data in terms of safety of the product. The stability profile can only be studied correctly when relevant and validated analytical methods are available.

Information from the production/purification processes

The development of the production and purification scheme is a slow and evolving process. Optimal production conditions and purification parameters can be defined only after several pilot scale production experiments. Each experimental batch has to be thoroughly studied and a large number of production parameters have to be measured, so as to compose a database, from which the history and evolution of the process can be reconstructed.

Some steps will have to be validated in the course of the development process. This applies especially to the elimination of certain impurities or also to validate the repeatability of a particular step. These validation data are highly important for later selection of specifications, and notably in choosing the purity testing. If validation data have clearly established that an impurity is effectively eliminated in the course of the process, it is entirely acceptable not to check for it routinely on each batch. Moreover, these validation data will also serve to identify the «critical steps», to which rigorous controls must be applied, and will allow In-Process controls and their limits to be defined and clearly justified. Thus for biotechnology products, the need for proper In Process controls should be emphasized.

Information from preclinical and clinical data

Information from preclinical and clinical data comes in parallel with the analytical development process. Once the first pilot batches have been produced, data acquired in animals or humans will add to the safety profile, notably in terms of tolerability and safety of use. The preclinical (and clinical) data obtained using the first batches will have to be studied in the light of the analytical data, and particularly the impurity profile of the given batch. It will sometimes be possible to correlate clinical activity with the results of tests in animals or cell cultures (bioassays).

Information from the experience gained during development and statistical analysis

Information from the experience gained and the statistical analysis is a notion probably even more difficult to analyse and adapt. However, it is important to note that if specifications are to be realistic and representative of what the manufacturer is capable, without being either too loose or too stringent, it is necessary to have a minimum of experience. A large database of results derived from a large number of tests must be compiled, even if it includes tests that will not be done routinely. The batch analysis results will be then analysed statistically to determine, parameter by parameter, the intrinsic variability of the production process and to fix, with a clear rationale, limits for the different specifications that will be applied routinely. In the statistical analysis, the performance of the analytical methods (accuracy, coefficient of variation, confidence limits, etc.) will also have to be taken in account when fixing the specifications.

Interim versus final specifications

It is clear that some of these sources of information are going to evolve in parallel, but at different rates. This means that specifications are also defined and redefined gradually as new information becomes available. It resembles a permanent feed-back loop in the course of which all new information can entail modifications of other aspects. For example, the impurity profile can only be studied correctly if (i) relevant analytical methods are used, and (ii) the purification process has been finalised.

Temporary (interim) specifications can be set, for application to the initial product used in clinical and preclinical trials. Similarly, on submission of the file, the proposed specifications can remain provisional, to be amended when sufficient experience has been obtained or industrial scale production has yielded a sufficient number of batches.

Thus, to fix specifications it is necessary to take into account data stemming from the development process and experience acquired. It will also be necessary to take into account the IPCS. The specifications for routine use are thus the end result of a rational selection process that establishes a fine balance between knowledge acquired through validation (these data are acquired once and for all), data that will be generated on each production batch through IPCS, and ultimate data generated by routine controls according to the chosen specifications.

Need for harmonisation

Currently, many biotechnology products are distributed world-wide. However, given the different requirements in the three regions, the manufacturer is obliged to repeat batteries of tests, which are all very similar, because particular methods or limits have been imposed in one region. This leads to a considerable loss of time, money and energy. It would be far better to harmonise the specifications so that more energy can be spent on quality control itself.

For the specifications concerning identity, purity and activity, harmonisation should essentially focus on definitions and on the way in which limits and specifications are fixed, steps that must include a statistical approach acceptable to all. This aim is reasonable, and the ICH can provide a suitable framework for such discussions. However, it seems difficult to harmonise an exhaustive list of test items for a given product. In Europe, each product «stands on its own merit» and it is up to the applicant to justify the choice of specifications and limits (even if the regulatory authorities sometimes tighten a given specification or add a test to the list). Whereas it will be difficult to harmonise on a given list of items, it is conceivable, however, to harmonise some methods that are common to all products, especially miscellaneous tests for pyrogens and sterility. It is illogical to repeat such tests simply because the methods and interpretation are different, when the aim is exactly the same. In this field, where most tests are «compendial», harmonisation could be based, for example, on the Pharmacopoeias.

Conclusion

In conclusion, specification setting is a very important step in qualifying the quality and safety of a product. Respect for specifications ensures that, regardless of the batch, the patient will receive a strictly equivalent product, despite the complexity of the production process or compound.

Several partners are involved in specification setting, from analytical development to industrial production. It is the responsibility of the applicant, using the described process and sources of information, to propose specifications that will satisfy the health authorities within the three regions.

Reference

1 Note For Guidance on production and quality control of medicinal products derived by recombinant DNA technology – EEC III/3477/92-EN – Revision Dec. 1994.

Dr. J.H. Trouvin, Agence du Médicament, Unité Biologie, 143 Bd. Anatole France, F-93285 Saint-Denis, France.

Brown F, Fernandez J (eds): Development of Specifications for Biotechnology Pharmaceutical Products. Dev Biol Stand. Basel, Karger, 1997, vol 91, pp 31-36

Specifications from a Biotechnology Industry Perspective

R.L. Garnick

Genentech, Inc., South San Francisco, CA, USA

Key words: Assay validation, potency, process validation, purity, strength, specifications, reference materials, reference standards.

Abstract: The emergence of new analytical technology and the production of pharmaceuticals for a global market in a cost-effective manner necessitate the establishment of worldwide specifications that are appropriate for the product and the manufacturing process. This requires a thorough knowledge of the protein and control of the systems that produce it as well as an understanding of the accuracy and precision of the assays used for testing. Harmonization of specifications among the worldwide regulatory authorities is critical for the future development of new pharmaceuticals. A continuing dialogue between industry and regulators to achieve this goal needs to be encouraged and supported.

INTRODUCTION

Setting appropriate specifications is one of the most difficult processes that a biotechnology company faces in the development of new pharmaceuticals. In the situation where the wrong specifications are set, the end result may be product recalls or the inability to produce a product. The future of the company may even be in jeopardy. Specifications for a biotechnology product should be based on a knowledge of both the protein and the processes used to manufacture it as well as on sound scientific principles. For example, the development of improved analytical methodology to detect lower impurity levels does not automatically mean that a specification should be changed or the process modified to remove the impurity. Not all impurities are deleterious or affect the efficacy of the product. Specifications may also need to be changed as additional information is gained about the protein. Initial specifications are usually based on a limited set of data from development runs and may need to be modified between the filing of the IND (Investigational New Drug) or CTX (Clinical Trial Exemption) and the NDA (New Drug Application), BLA (Biologics License Application), or MAA (Marketing Authorization Application).

This paper provides an overview of how specification development has changed in the biotechnology industry, what some of the issues are currently in setting specifications, and what parameters should be considered. Some specific examples of specifications for identity, purity, potency, and strength are also included.

Historical Perspective

The scope of specifications for biotechnology products has changed since the inception of the industry in the late 1970s. From this time period until about 1985, the industry and regulatory authorities took a very cautious position on biomolecules from this new and emerging technology. It was not possible to characterize these molecules thoroughly in an analytical way, so the process of manufacture was critical in defining the product. This made manufacturing changes difficult to implement. There was major concern from an immunological perspective about these new products and their associated impurities, so specifications were rigorously developed to assuage the regulatory authorities. There was the belief that the purer the product, the fewer the regulatory hurdles that needed to be overcome. This resulted in increased expenses for product manufacture and redundant testing.

During the next decade (1985-95), there was an explosive growth in the analysis of proteins. Techniques such as peptide mapping/mass spectrometry, capillary electrophoresis, carbohydrate analysis, high-performance liquid chromatography (HPLC), and high precision biomimetic cell line-based bioassays were shown to be of great value in the analysis of biopharmaceuticals. Transgenic animal models were developed to predict immunogenicity, and there were positive clinical experiences with products such as tissue plasminogen activator and erythropoietin and their impurities, leading to reduced regulatory concerns. In addition, the focus shifted from a domestic base to global drug development and the implementation of multiuse facilities for multiple product manufacture.

During the past year (1995-96), the concept of well-characterized proteins has evolved, resulting in less of a reliance on the manufacturing facility and consistency of the process for drug production and a broadening of the application of the principles of process validation. Because of the expanding global marketing of biopharmaceutical products, the need for worldwide harmonization of specifications and regulatory requirements is of paramount importance to the biotechnology industry. An attempt to meet this need is being made by the International Conference on Harmonisation (ICH), which is currently working on guidelines for worldwide specifications.

Definition

Specifications can be understood as a group of requirements that define certain critical attributes of a material based on predefined values. They represent a set of standards against which a product is judged to be acceptable or not. Specifications are necessary to meet good manufacturing practices (GMP) requirements and they serve as a tool to demonstrate consistency of production. The scope of specifications critical for the biotechnology industry encompasses raw materials, in-process controls, the bulk drug substance, the final product, and final product stability. All these specifications together allow a manufacturer to define what it is that the product must represent consistently from lot to lot.

Current Issues Related to Specifications

Conflicting standards worldwide are a great impediment to the development of new pharmaceuticals. For example, the test for sterility differs between the U.S., Europe, Japan, and Australia. In addition, it is usually necessary to change specifications between the development of a new drug and its final product formulation. A specification appropriate for a drug in development is usually revised by the final product stage and may need to be changed again after product approval to incorporate improvements in the manufacturing process. The mechanisms associated with making these changes can be very cumbersome and lengthy. There is a tension between the development companies, who want to minimize the costs of testing their products until they are shown to be effective, and the regulators, who are concerned about the safety and efficacy of these products. The cost of testing must be balanced against the value of the results from these tests. Finally, there is the important relationship between the use of process validation and the value of testing.

Process validation is the demonstration through a proactive validation plan of the ability of the process to perform its tasks reliably. It is generally more cost-effective and efficient to use process validation rather than to set specifications and test every lot to determine if it meets those specifications. Some examples of the use of process validation include viral removal and inactivation, DNA removal, host cell impurity removal, and small molecule removal.

Specifications should be used wherever a critical parameter cannot be validated or when it is necessary to define a critical performance criterion for a product or intermediate. For example, if it is known that a particular protein is subject to deamidation and this is of concern, it may be appropriate to set an in-process specification for deamidation. Whenever possible, in-process controls, which are evaluated either by statistical methods or action limits, should be used in place of specifications during the production of the material. These controls provide monitoring of how a specific step in the process is performing and allow for corrective action in time to avoid failing specifications at the bulk drug or final product stage. These limits should be reviewed and reset periodically, if appropriate.

An example of a control chart is shown in Fig. 1. A chart such as this enables the tracking of a process or test over time and will readily show an outlier, alerting appropriate personnel that something may be wrong with the process/product or the method.

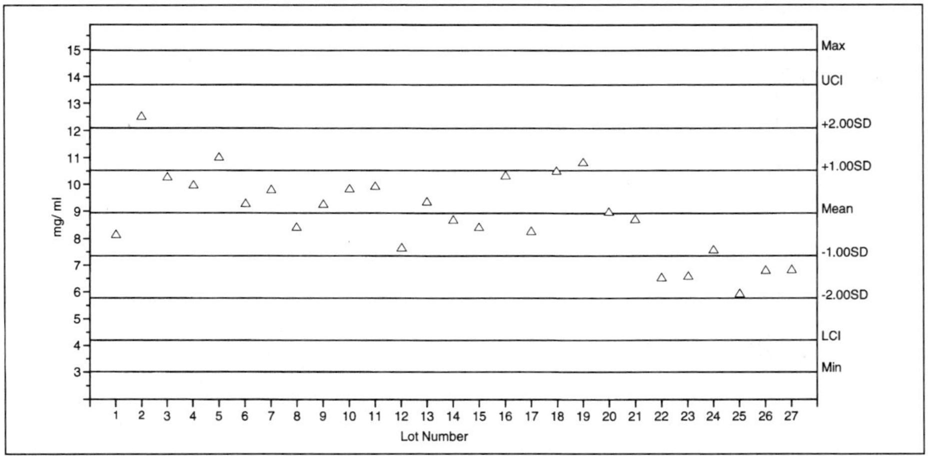

Fig. 1: Control chart for a protein concentration assay showing the upper and lower action limits (called Max and Min, respectively). The UCI and the LCI are the upper and lower confidence intervals, respectively.

Setting Specifications

Some specifications are provided in various compendia, such as the Pharmacopoeias of the U.S., Europe, and Japan, or in the U.S. Code of Federal Regulations (CFR). These include tests for sterility, colour, appearance, and clarity, moisture, and mycoplasma. However, most specifications are set based on knowledge of the specific product gained from clinical experience, an understanding of the purity of the product that the process can deliver, and the effect of impurities on the efficacy and safety of the product. Many specifications are set based on simple statistics, such as two or three standard deviations about the mean from historical production values or fiducial limits for potency dictated by regulatory requirements. Lastly, specifications may be set based on company negotiations with regulatory authorities.

Specifications should be set based on validated methods. It is difficult to set limits on any given parameter if it is not known how «good» an assay is in evaluating that parameter. To obtain the correct value for an analyte, the number of replicate analyses required must be determined during the assay validation process. Both the accuracy and precision of an assay must be considered. As shown in Fig. 2, some assays are both accurate and precise, some are accurate but imprecise, some are precise but inaccurate, and some are both imprecise and inaccurate. These latter assays should not be used to establish specifications.

Two-Tiered Specifications

Most protein products degrade with time and this fact must be considered in the design of specifications. To control for protein degradation over the shelf life of the

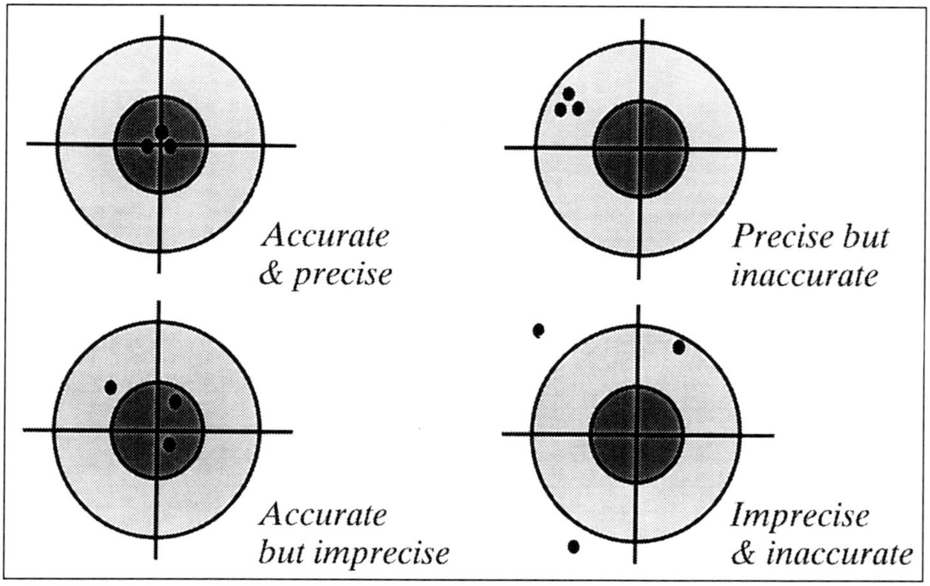

Fig. 2: Examples of the meaning of accuracy and precision in assays.

product, two-tiered specifications are desirable. This means one specification is set for the time of release of the product, and another is set for the product at the end of its shelf life. However, these types of specifications require that the manufacturer thoroughly understands the degradation profile of its products and can justify the two sets of specifications.

Specific Examples of Specifications

Specifications are typically used for controlling the identity, purity, potency, and strength of the product. For example, the specification for identity is generally «positive identity», or «consistent with» a standard or reference material. The analytical methods typically used include peptide mapping, electrophoresis, immunological assays, HPLC, and enzyme substrate reactions. For a purity specification, the question of how pure does a biological need to be must be addressed. Experience has shown that many product-related impurities are equipotent and non-immunogenic. Some common product-related impurities and the specifications set for them are shown in Table 1.

Methods for determining the potency of biomolecules are the only ones that provide information on the true secondary and tertiary structures of a molecule. These methods answer the question of whether or not the molecule works for its designated purpose. Specifications commonly set for potency based on the four types of assays currently available are shown in Table 2.

Table 1: Specifications commonly set for product-related impurities.

Product-related impurities	Specifications used by industry
Aggregates/clips (by Molecular size distribution)	> 95% monomer
Aggregates/clips (by SDS-PAGE/silver stain)	No significant new bands not present in reference material that are greater than that of bovine serum albumin (BSA) control
Oxidized forms (by HPLC)	Not more than 10-15%
Deamidated forms (by HPLC)	Not more than 5-30%
Norleucine level (by HPLC)	Not more than 2%

Table 2: Specifications for potency.

Assay type	Specification target
Animal-based bioassays	70-130% of label claim
Cell line-based bioassays	80-120% of label claim
In vitro enzymology assays	85-115% of label claim
Receptor binding assays	85-115% of label claim

The strength or protein content usually has a specification based on compendial articles (e.g., the Pharmacopoeias), and is typically set at 90-110% of label claim. The lower limit is critical, based on legal concerns; a product can be considered to be adulterated if it cannot meet this specification. The specification may be set based on a number of appropriate assays, such as UV spectroscopy, HPLC, quantitative amino acid analysis, or colorimetric or redox methods (e.g., Lowry/Bradford assays). The UV spectroscopy method has the advantage of being standard-independent; the other methods all require the use of a reference standard or a reference material. The specification is usually expressed in international units (IU)/mL or mg/mL for liquid products and in IU/vial or mg/vial for lyophilized products.

Standards and Reference Materials

No discussion of specifications is complete without including standards and reference materials. There is often confusion because of inconsistency in the nomenclature used. A primary reference standard is internationally recognized and obtainable from pharmacopoeial organizations (e.g., U.S. and Europe), the World Health Organization (WHO), or the Food and Drug Administration (FDA). A secondary standard has been calibrated against a recognized primary standard. A reference material is a thoroughly characterized preparation that is used for comparative purposes. It has not been referenced against a standard. To set appropriate specifications, it is critical that there be proper selection and characterization of reference standards and reference materials.

Conclusions

The global pharmaceutical industry faces great challenges in the development of new and improved pharmaceuticals in an ever changing world market and regulatory environment. On the other hand, a unique opportunity exists among the various regulatory authorities and industry to harmonize conflicting and redundant requirements to facilitate the development and approval of new pharmaceuticals in a cost-effective manner. To further this goal, the pharmaceutical industry recommends the following: (a) that one set of specifications be accepted worldwide by regulatory agencies; (b) that specifications be rational and appropriate; (c) that specifications be set based on sound scientific principles; and (d) that mechanisms be established for resetting specifications when necessary and appropriate.

Dr. R.L. Garnick, Genentech, Inc., 1 DNA Way, South San Francisco, CA 94080, USA.

Brown F, Fernandez J (eds): Development of Specifications for Biotechnology Pharmaceutical Products. Dev Biol Stand. Basel, Karger, 1997, vol 91, p 37

SESSION II

Evolution of Specifications Through Product Development

Moderator: *Vincent Anicetti, M.S.*

Brown F, Fernandez J (eds): Development of Specifications for Biotechnology Pharmaceutical Products. Dev Biol Stand. Basel, Karger, 1997, vol 91, pp 39-43

Appropriate Specifications at the IND Stage

J. Geigert

Quality, IDEC Pharmaceuticals Corporation, San Diego, CA, USA

Abstract: Biopharmaceutical product specifications are to be «scientifically sound and appropriate». However, how does a biopharmaceutical company determine what are acceptable specifications for its product at the early stages of clinical development? Are there really enough test data at the start of the IND trials to «establish» specifications? As an industry, are we doing too many tests at the IND stage (i.e. following the «must do every test on someone's published list» syndrome), and then finding that specifications need to be set for each test we run? Are we feeling pressurised (either by regulatory agencies or by our own corporate cultures) to set unrealistic and tight specifications at the IND stage, and then later regretting how to justify loosening them when we gain more experience with the release testing or stability of our products? A discussion on how to set practical product specifications for biopharmaceutical products at the IND stage is presented.

Discussion

A rigorous approach is needed to ensure the quality of large, complex biopharmaceutical products. Adequate in-process testing, thorough process validation, and a comprehensive assessment of final product quality using a battery of analytical methodologies, are all needed to demonstrate lot-to-lot consistency. The virtual explosion in the availability of sophisticated and reliable analytical instrumentation, as well as extensive and sensitive chromatographic, immunological and spectrometric methods of product analysis, continue to extend the range of the definition of «well-characterized» for biopharmaceuticals.

Under current regulations, any use in the United States of a drug product not currently authorized for marketing in the United States first requires submission of an Investigational New Drug (IND) application to the FDA. The start of the human clinical trials is referred to as Phase 1, with the primary purpose of assessing safety of the drug product. Guidance from the FDA has been limited on what specifications are required at the early stages of clinical trials. In the United States Code of Federal Regulations on Good Manufacturing Practice [1] the following statement is made:

> «Laboratory controls shall include the establishment of scientifically sound and appropriate specifications, standards, sam-

pling plans, and test procedures designed to assure that components, drug product containers, closures, in-process materials, labeling, and drug products conform to appropriate standards of identity, strength, quality, and purity.»

While it remains for the individual company to define 'scientifically sound and appropriate' for their specific protein product, the FDA has recognized that the specifications at Phase 1 may be minimal.

FDA's expectations for specifications at the Phase 1 stage have recently been formalized in a Guidance Document for Industry [2]. The following are the stated acceptable limits and analytical methods to be used to assure the identity, strength, quality and purity of the Phase 1 products:

Drug Substance (Bulk Drug)

«A brief description of the test methods used should be submitted. Proposed acceptable limits supported by simple analytical data, (e.g., IR spectrum to prove the identity, and HPLC chromatograms to support the purity level and impurities profile) of the clinical trials material should be provided. Submission of a copy of the certificate of analysis for the drug substance used in the clinical batch is also suggested. The specific methods will depend on the source and type of drug substance (e.g., animal source, plant extract, radiopharmaceutic, other biotechnology-derived products). Validation data and established specifications ordinarily need not be submitted at the initial stage of drug development. However, for some well-characterized, therapeutic biotechnology-derived products, preliminary specifications and additional validation data may be needed in certain circumstances to ensure safety in Phase 1.»

Drug Product (Finished Product)

«A brief description of the proposed acceptable limits and the test methods used should be submitted. Tests that should be submitted will vary according to the dosage form. For example, for sterile products, sterility and pyrogenicity tests (or endotoxin assays, as appropriate) should be submitted. Submission of a copy of the certificate of analysis of the clinical batch is also suggested. Validation data and established specifications need not be submitted at the initial stage of drug development. For well-characterized, therapeutic, biotechnology-derived products, adequate assessment of bioactivity and preliminary specifications should be available.»

Preliminary specifications are acceptable at this early stage of clinical development, and this FDA approach makes good common sense. Since companies today are expediting the entry of their products into clinical programmes, they are now finding that, at the time of filing the IND, they have less experience, both with the control of the manufacturing process and with the release testing, than previously. Typically today, companies may file their IND with three or fewer lots. Therefore, preliminary specifications at the IND filing, with further evaluation and analysis over the clinical development programme, eventually leading to established specifications by the time of filing the PLA or NDA (see Fig. 1), works to everyone's advantage.

Taking full advantage of this Guidance Document for Industry requires a corporate culture shift. It was less than two years ago that my former company filed an IND for a new therapeutic, biotechnology-derived protein product. Within three months of filing the Phase 1 IND, we received a six-page, 20-question letter from CBER. The very high expectation by that reviewer of the level of established specifications required at Phase 1 was obvious from the comments:

> «Please quantify/identify the number of bands present and set specifications for the multiple bands present in the isoelectric focusing.»
>
> «With regard to the carbohydrate analysis that is in progress, the sites of glycosylation, the number of glycoforms, etc., need to be determined and specifications need to be set for major glycoforms.»
>
> «Please set specifications with regard to an acceptable range for the bioassay. Information should be provided regarding the reproducibility, reliability and precision of this assay. The WHO International Standard should be used to standardize this assay.»

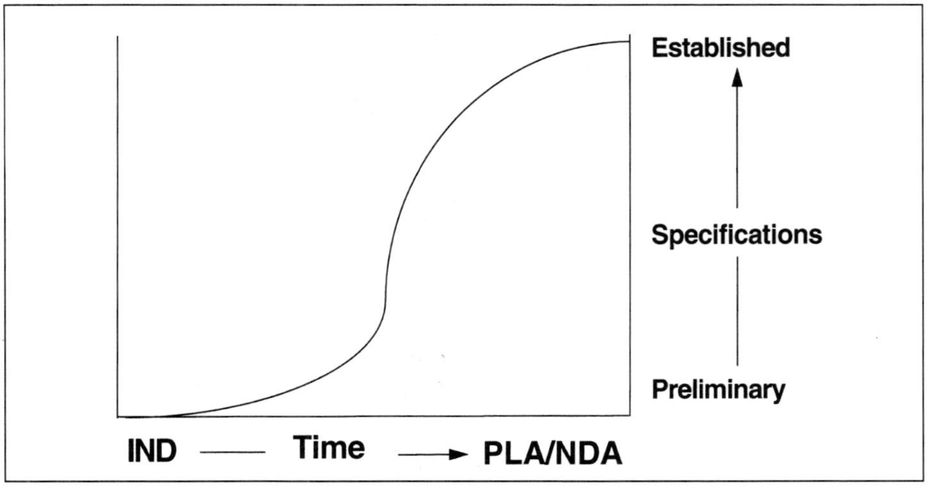

Fig. 1: Development of a specification over the course of clinical development from the filing of the IND to the filing of the Product License Application (PLA) or the New Drug Application (NDA).

In receiving letters such as this, it is not surprising that a company would consider running every available test method and setting a tight specification for each, even though it might not result in a better quality product. After all, it is more palatable to do this than receive such an embarrassing FDA letter. But times change and so corporate culture also needs to change. Figure 2 demonstrates how a company can take advantage of the recent Guidance for Industry in reducing the number of tests to run, which results in less specifications being needed.

What is an acceptable «preliminary specification» for a Phase 1 product? As illustrated in Figure 3, it might be a common sense approach such as «report result»

	Product A IND filed in 1994	Product B IND filed in 1996
Identity/ Integrity	7	2
Purity/ Impurities	3	2
Potency/ Strength	2	2
Safety/ General quality	4	4
Total:	16	10

Fig. 2: Change in the number of specifications required for release of a therapeutic, biotechnology-derived bulk drug protein product at Phase 1, between the years 1994 and 1996.

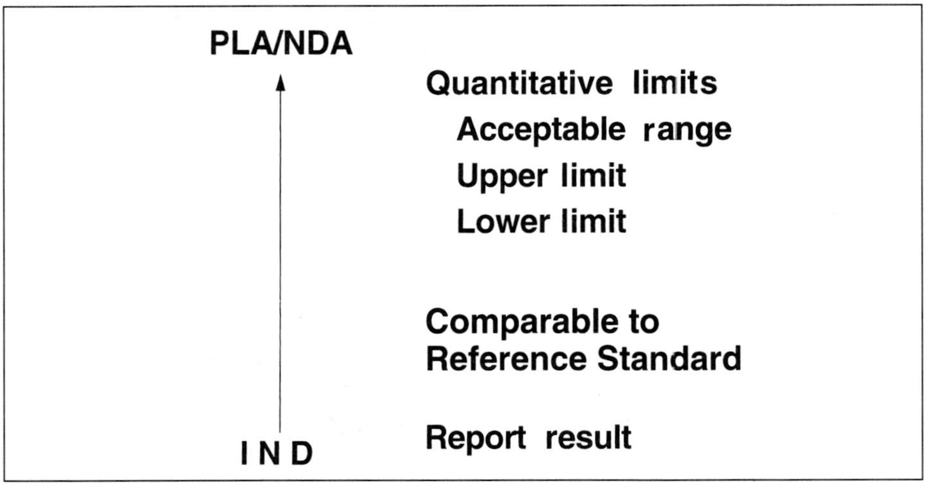

Fig. 3: Illustration of the evolution of a preliminary specification to an established specification over the course of clinical development.

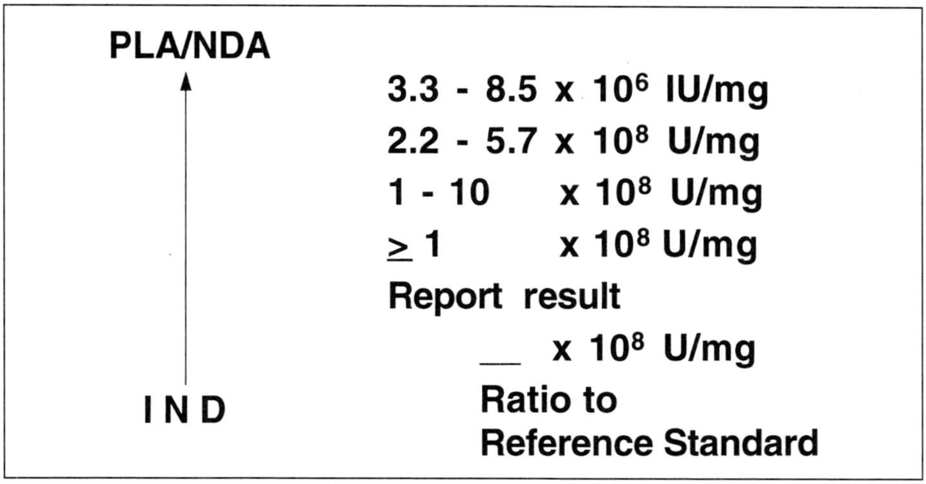

Fig. 4: Illustration of the evolution of an adequate assessment of bioactivity over the course of clinical development.

or «comparable to the Reference Standard». With greater experience with the product through the clinical development programme, eventually a more statistical approach would become appropriate, and lead to an established specification. It should be pointed out that preliminary specifications should not be applied to safety parameters. All regulatory limits, such as LAL endotoxin and/or rabbit pyrogen, sterility and general animal safety testing, must be met.

What is an «adequate assessment of bioactivity» for a Phase 1 product? As presented in Figure 4, it could be just «report result», either as a reported specific activity or a ratio to the Reference Standard. With greater experience with the product through the clinical development programme, an emphasis on statistical analysis would occur, most probably starting with a lower limit, then setting a range, followed by a tightening of this range. Finally, the specification would be established in terms of International Units (IU).

With all the emphasis currently on FDA reform, opportunities for adjusting the required specifications at the early IND stages should continue to occur. It is the responsibilty of the industry to undergo the corporate culture shift to take advantage of these opportunities.

References

1 United States Title 21 Code of Federal Regulations, Part 211. Current Good Manufacturing Practice for Finished Pharmaceuticals.

2 FDA Guidance for Industry, 1995: Content and Format for Investigational New Drug Applications (INDs) for Phase 1 Studies of Drugs, Including Well-Characterized, Therapeutic, Biotechnology-Derived Products.

Dr. J. Geigert, IDEC Pharmaceuticals Corporation, 11011 Torreyana Road, San Diego, CA 92121, USA.

Brown F, Fernandez J (eds): Development of Specifications for Biotechnology Pharmaceutical Products. Dev Biol Stand. Basel, Karger, 1997, vol 91, pp 45-48

Global Regulatory Considerations for Unified Assay Specifications

S. Patterson, L. Lin, J. Elchico, R. Fuchs, S. Powell, D. Redhair, M. Warwick, N. Solli, B. Méndez

Chiron Corporation, Emeryville CA, USA

Abstract: When developing a biotechnology product for global registration, there are several aspects to evaluate in an effort to unify specifications. These include differences between the United States, Europe and Japan, and Rest-of-World (ROW) countries with regard to the respective regulatory guidelines and pharmacopoeias in force, the state-of-the-art of product testing analytical methods, and the interval between submitting a registration dossier to different countries. In terms of regulatory guidelines, one country may have a monograph or required specifications for particular tests, for example the potency that a product has to meet before clinical trials can be initiated. For pharmacopoeias, different assay methods are required for sterility, general safety, and pyrogen testing, so that one may have to test a specific lot of a product at two or three different times to evaluate the same parameter, because of specific testing differences required for each country's pharmacopoeia. In addition, the state of analytical methods is always evolving and better analytical techniques become available. Sometimes, from starting with one set of tests, and based on the time in development, new tests may be added to the existing list of release specifications, because new analytical techniques have become available.

Examining the global registration approval process for Betaseron®, (interferon beta-1b) illustrates when specifications were able to be unified and when they were not.

Introduction

This paper addresses the various aspects of product testing and specifications to consider and evaluate when a recombinant protein is being developed to obtain product registration worldwide. For biotechnology products, there is no master list of tests and limits for product specifications; however, even for products that share a similar profile, each product needs to be evaluated on a case-by-case basis, both internally and by the various regulatory authorities. Betaseron, one of the first biotech drugs approved on a global scale, is presented here as a test case for discussion of global regulatory considerations for attempting to obtain one set of specifications.

Materials and Methods

Betaseron or interferon beta-1b is a single polypeptide chain analogue of human interferon beta with the following properties: the amino terminal methionine is removed during production, meaning that it has 165 amino acids; serine is substituted for the odd cysteine at position 17 to prevent incorrect disulphide bond formation. Interferon beta-1b is expressed in *E. coli* under the control of the *trp* promoter but due to its manufacture in *E. coli* it is non-glycosylated. Other than these described differences, the amino acid sequence of Betaseron is identical to that of native IFN-ß with a single disulphide bond between the cysteine residues at positions 31 and 141.

Interferon beta-1b was cloned and expressed in *E. Coli* in 1981. An interferon beta mutein, where the cysteine at position 17 was replaced by serine, was constructed using oligonucleotide-directed site-specific mutagenesis [1]. The resultant interferon beta mutein, IFN-ß Ser 17, was further developed and an Investigational New Drug application (IND) was filed in April 1983 to investigate its use as a cancer agent. Various clinical trials were conducted in a number of clinical indications over the next four years, one of which included evaluating its effect in multiple sclerosis patients. Pivotal clinical trials of Betaseron in Relapsing-Remitting Multiple Sclerosis began in 1987 and the first registration dossier, the U.S. PLA/ELA, was filed in July 1992. The FDA licensed Betaseron within one year [2]. It should be noted that development of Betaseron was a joint venture of Chiron and Schering AG, the parent company of Berlex Laboratories, Inc. Schering AG submitted an application for marketing authorization or MAA to the EMEA in May 1994. The EMEA, which stands for the European Agency for the Evaluation of Medicinal Products, is the European Union regulatory body that came into effect in 1995 and comprises fifteen European countries. At the time of filing in 1994, Betaseron was filed to the twelve European Union Member states under the concertation procedure. (This was converted to the centralized procedure established on January 16, 1995). Eleven other or ROW countries, in addition to the fifteen European countries in the EU, received the Betaseron registration dossier in 1994.

The product received EMEA approval in November 1995 [3]. Additional approvals were received in 1995 in Australia, Canada, Israel, and some Eastern European countries. Clinicals trials are in progress in Japan and a request for marketing authorization will occur upon completion of those clinical trials. Additional approvals were received in 1996 from countries in South America.

Results

The Betaseron manufacturing process consists of fermentation, recovery, a series of purification steps, and formulation. The purified interferon beta-1b protein is formulated with human serum albumin (HA), dextrose, and lyophilized. FDA classified Betaseron interferon beta-1b as a well-characterized protein in 1995. The primary sequence, secondary structure, and tertiary structure of this protein have all been identified and defined.

The complete amino acid sequence of Betaseron was verified by amino and carboxyl termination sequence analysis, peptide mapping after endoproteinase Lys-C digestion, fragment analysis after chemical cleavage of methionine and at tryptophan, and mass spectrometry. Overlapping amino acid sequence information obtained confirmed that the amino acid sequence is the same as predicated by the DNA sequence. The amino-terminal methionine of IFN-β also serves as the initiation codon for production in *E. coli* and is removed after synthesis by the *E. coli* enzyme methionine amino peptidase, resulting in a 165 amino acid protein with a homogeneous N-terminus.

The secondary structure was confirmed by using both peptide mapping and circular dichroism. An intramolecular disulphide bond between Cys-31 and Cys-141 is confirmed by peptide map analysis. Using CD, the molecule is found to be approximately 45% alpha helix, with the rest being beta-sheets, a structure quite similar to that found for native interferon beta.

The tertiary structure was evaluated using a potency assay. Bioactivity was measured by antiviral activity in vitro using a cytopathic effect assay. Briefly, the assay measures the protection to viral challenge induced by exogenously added IFN-β to susceptible indicator cells. Titre is defined as the dilution which induces protection in 50% of the cells. The assay is calibrated against the WHO recombinant IFN-β reference.

Each lot of Betaseron is tested for identity, purity, strength, potency, safety and quality. There are 46 different tests with a total of 60 release specifications for each lot of Betaseron. These include testing from the master cell bank, through to bulk drug substance, and finally to the final product.

The identity of Betaseron is confirmed by Lys-C peptide mapping, N-terminal sequencing, C-terminal sequencing, and amino acid analysis. The purity of Betaseron is analysed by SDS-PAGE Fast Green, non-reducing and reducing SDS-PAGE, and reversed-phase HPLC analysis.

Potency is determined by the CPE bioassay in vitro. Strength is determined by IFN-ELISA and total protein by Lowry. Safety is evaluated by several tests: determination of *E. coli* antigens, plasmid DNA, sterility and general safety, LAL, bioburden and pyrogen.

The quality of Betaseron is determined by testing for pH, appearance, moisture, osmolality (salt content), human serum albumin, a stabilizing agent used during formulation, dextrose, and SDS.

Discussion

Due to different requirements for the number of identity tests, product purity levels, specific potency-testing calculations, types of safety testing, and quality tests, the Betaseron commercial product has seven different Certificates of Analysis (C of A) to date: U.S., Canada, Europe, Australia, Switzerland, and Rest of World. Japan has a separate C of A; the product is currently in clinical trials in Japan and not approved for commercial use.

Most countries require a minimum of two identity tests for the product, usually a peptide map and N-terminal amino acid sequencing. One country has required two additional tests, a C-terminal assay and an amino acid analysis. I do not believe that these additional tests increase the knowledge of the identity of Betaseron. In fact, for all the lots tested over the years, there has never been a discrepancy in any of the identity tests conducted. For purity tests, we have been required to increase the purity limit by both SDS-PAGE and reverse-phase HPLC for some countries even though the product was commercially available in the United States at the time the registration dossier in those countries was under review.

For potency testing, the in vitro CPE bioassay, is conducted for each country, but the method of calculation varies. Also, one country requires testing of potency at both the formulated bulk and final product stage, while other countries require fiducial specification limits.

Each country has slightly different procedures required for testing the sterility, general safety and pyrogenicity of the product. This means that each lot may have to be tested in different ways to evaluate the same parameter. For instance, for sterility testing, the methodology or test duration is different in different countries. This requires redundant testing of the same lot. In the U.S., sterility testing is a seven-day test, and is required to be conducted on both the formulated bulk and final vials. In Europe, the testing is slightly different from that in the U.S. In Australia, sterility testing is a 14-day test. Testing a lot three different times for sterility dramatically increases the risk of rejecting a lot unnecessarily. Sterility testing requires extreme care and precision to avoid false-positive results. Due to the market requirements in different countries, one cannot always designate an entire lot for one country and duplicate sterility testing must be carried out.

Government lot-release testing following Company lot-release testing is another concern. It delays the use of the product, reduces the time between placing the product into commercial distribution and the end of the shelf life of the product. FDA eliminated this requirement in November 1995, when it designated Betaseron a well-characterized protein. However, Canada maintains lot-release.

In summary, Betaseron is a 165 amino acid single polypeptide chain, with one intramolecular disulphide bond, and is non-glycosylated. But the requirements placed on the product from fermentation to final product include a total of 46 release tests and 60 release specifications. This requirement is a significant level of redundant testing. Differences between specifications among different countries has required use of six different commercial Certificates of Analysis.

The following recommendations are for consideration during International Harmonization discussions: (i) reduce the number of identity tests to one or two, such as peptide mapping and N-terminal analysis, and eliminate the need to carry out C-terminal analysis and amino acid analysis in QC settings. (ii) standardize the purity limit for all countries. (iii) unify the calculation of potency among countries, i.e. test for potency in the final product only and not in the formulated bulk. (iv) waive the release testing for residual host contaminants, such as *E. coli* antigen and plasmid DNA, based on process validation data demonstrating an adequate removal of each during the manufacturing process. (v) unify methodology for the specification of assays from various pharmacopoeias for sterility testing, pyrogen testing, and general safety. (vi) eliminate government-required lot-release testing for all countries. (vii) use the registration review period to test qualification lots and, based upon confirmation of the Company's results, eliminate routine testing on a lot-by-lot basis.

Conclusion

When developing specifications for biotechnology products, there are a number of complex issues to consider and evaluate. The goal should be to unify both test methodology and specifications as far as possible, so that when a product is distributed worldwide, it can meet one set of tests and limits acceptable for use in any country that has approved its use commercially. This would allow one Certificate of Analysis with an appropriate number of identity, purity, potency, strength, and quality tests to ensure that all product requirements are met.

References

1 Mark D, Drummond R. Creasey A et al: A synthetic mutant of interferon beta for clinical trial, in Kishida T (ed): Proc International Symposium on Interferons, Kyoto, 1984, pp 167-172.

2 U.S. Summary Basis of Approval July 1993.

3 European Public Assessment Report, Betaferon, November 1995.

Dr. S. Patterson, Chiron Corporation, 4560 Horton Street, Emeryville, CA 94608, USA.

Brown F, Fernandez J (eds): Development of Specifications for Biotechnology Pharmaceutical Products. Dev Biol Stand. Basel, Karger, 1997, vol 91, pp 49-54

Characterization and Establishment of Specifications for Biopharmaceuticals

G.C. Davis, R.M. Riggin

Lilly Research Laboratories, Eli Lilly and Company, Lilly Corporate Center, Indianapolis IN, USA

Abstract: This paper describes the role of product characterization and product specifications in the context of an overall control strategy. Due to advances in analytical characterization and biotechnology manufacturing methods, product characterization can now be used as an effective means of assessing the impact of manufracturing process changes on product quality, safety and efficacy, hence obviating the need for repeating clinical testing each time a manufacturing process change is made. The ICH specifications guidance document can serve a key role in harmonising test requirements and regulatory practices with regard to produce characterization. Consumers, regulatory agencies, and biopharmaceutical manufacturers all stand to benefit from this harmonization process.

Introduction

Historically many biological pharmaceutical products have been derived from natural sources and frequently the therapeutic agent has been difficult to characterize in detail with available analytical methods. Under those circumstances, control of the product has required strict definition and careful control of the production process to ensure product quality, safety and efficacy. Since only limited characterization was possible, significant changes in the manufacturing process needed additional clinical studies to ensure that product quality, safety and efficacy were maintained. However, advances have been made over the past 10-15 years in both manufacturing technology and assay methodology. Recombinant DNA production methods have been much more routine and better-understood. Structural characterization methods such as electrospray mass spectrometry (ESMS) and matrix-assisted laser desorption mass spectrometry (MALDI) enable researchers to characterize complex proteins, including many glycoproteins, in great detail. In view of these technological advances, the impact of a change in the manufacturing process for a well-characterized biopharmaceutical on product quality, safety and efficacy can usually be assessed by physical, chemical and biological characterization, without the need to conduct additional clinical studies. Obviously a key consideration in assessing the potential impact of manufacturing changes on product quality, safety and efficacy is the characterization dataset available for the particular product, i.e. when is a product well-characterized? This paper describes some considerations in selecting appropriate characterization strategies for biopharmaceuticals.

The establishment of specifications for biopharmaceutical products is a topic of great interest to both regulatory officials and the biopharmaceutical industry. Efforts are being made (within ICH) to harmonize specifications on a worldwide basis. These efforts are important since the biopharmaceutical industry is increasingly a global industry and a uniform policy for setting specifications will streamline product development and regulatory review processes, thus benefitting industry, regulatory authorities and the consumer. This paper describes the purpose of product specifications in the context of an overall control strategy and will discuss some of the considerations in harmonizing specifications on a worldwide basis.

An Overall Control Strategy

When discussing control of a product to ensure its quality, safety and efficacy, one usually focusses on the product specifications. However, product specifications should be viewed as only one element in an overall control strategy. Table 1 lists the elements which are normally part of the control strategy for a biopharmaceutical product. The control strategy includes development information collected from a process fully representative of the commercial scale process (usually this information is collected at pilot-scale). Process validation data are collected to confirm the consistent removal of key impurities (e.g. host-cell proteins, host cell DNA, adventitious agents). The drug substance in fully characterized using appropriate physicochemical and biological/biochemical assays. Additional characterization assays for the drug product may be necessary, in cases where the formulation, filling, or finishing processes can affect the phyicochemical or biological/biochemical properties. Drug substance and drug product stability are evaluated in accordance with ICH guidelines [1]. The stability information is used in the context of the overall control strategy for two major purposes: (i) stress stability studies are used to elucidate the degradation pathways of the drug substance, thereby allowing selection of appropriate stability-indicating assays and (ii) degradation rates for the drug product at the intended storage condition are used in the determination of appropriate release

Table 1: Elements in an overall control strategy.

Development information – Process validation – Drug substance/product characterization – Drug substance/product stability evaluation
Raw material controls
In-process controls – In-process assays – Control of critical process parameters (e.g. pH, hold times, etc.)
Bulk drug substance release assays (specification testing)
Final drug product release assays (specification testing)

and shelf-life specifications. The remaining components in the control strategy are raw material controls, in-process controls, bulk drug substance release assays and final drug product release assays (product specification testing).

To determine the appropriateness of proposed product specifications those specifications must be considered in the context of the overall control strategy and available development information for that product. The guiding principle in assessing a particular set of proposed product specifications should be to determine whether or not the overall control strategy achieves the requirement of assuring that the product will be consistently produced with an acceptable degree of quality. The optimum control strategy will vary from one component, product or process to another, hence a high degree of flexibility in selection of a control strategy is needed. For example, in cases where process validation data are provided, demonstrating removal of a process ingredient to an acceptable level, there should not generally be a need to perform lot-release testing for that component in the product. Conversely, for a substance such as endotoxin (in an injectable preparation), there is a clear need for lot-release testing since endotoxin could potentially be introduced into the product at almost any stage of processing, and the level of endotoxin can affect patient safety. In some instances it may be advantageous to perform a particular assay on the drug substance, rather than on the drug product. However, such an approach is appropriate as long as the property determined for the drug substance is not affected by the formulation, filling, or finishing processes. Similarly, in-process testing may be more effective than lot-release testing in controlling certain contaminants, as long as the contaminant is not introduced into the process downstream of the in-process test point.

The remainder of this paper focusses on two key elements in the control strategy, drug substance and drug-product characterization and specifications.

Drug Substance and Product Characterization

The major objectives of characterization studies are as follows: (i) to provide a comprehensive understanding of the chemical structure, physical properties, biological/biochemical properties, impurity profile and degradation pathways of the drug substance, (ii) to serve as a guide in selecting specification assays and stability-indicating assays and (iii) to determine the effect of manufacturing process changes on the drug substance and/or drug product. The elements typically included in a characterization package are shown in Table 2. The complete characterization package is usually performed on a single batch (often designated as the primary reference standard), although some aspects (e.g. the impurity profile), as well as a complete battery of specification assays, will need to be performed on several batches which are representative of the commercial manufacturing process. Some elements of the characterization package should also be used to assess product comparability to support manufacturing process changes [2]. The specific assays performed for a given product should be selected in view of the structural and biological properties of the drug substance.

Biological products which have been characterized to a high degree of specificity are deemed «well characterized». An important attribute of well characterized biological products is the ability to determine the effect of process changes on product quality, safety and efficacy, without the need to perform additional clinical studies. Since process changes are often required in the pharmaceutical industry, including the biopharmaceutical industry, to keep pace with technology it is important to define when a product is well characterized, thereby allowing process changes to be implimented expediously. At an FDA meeting in December, 1995, a preliminary definition for well-characterized proteins was presented by the biopharmaceutical industry [3]. The essential elements of this definition are shown in Tables 3 and 4. The principles and definitions provided in these tables can serve as a useful guide in constructing an effective characterization plan and overall control strategy for the product.

Table 2: Elements typically included in a characterization package.

Determination of amino acid sequence
Evaluation of higher order structure (including disulphide bonds)
Evaluation of biological/biochemical properties (especially biopotency/specific activity)
Evaluation of post-translational modifications (microheterogeneity)
Evaluation of the impurity profile
Determination of quantity (e.g. protein content) – mass balance, if feasible
Evaluation of stability under stressed conditions to define degradation pathways

Table 3: Proposed manufacturing principles to be applied to the manufacturing and testing of well characterized biopharmaceutical products.

The manufacturing process should be robust, reproducible, validated and designed to produce the active drug substance.
The major toxicity should be associated with the intrinsic biological activity of the drug, not impurities.
Process validation further ensures safety.
Tests for identity, purity, potency and strength should be sensitive, quantitative and validated.
Specifications should be established based on manufacturing history and clinical experience. Many process-related impurities (e.g. host cell proteins and DNA) should be controlled by process removal validation, in lieu of lot-release testing.

Table 4: Proposed definition of a well characterized protein.

Identity Amino acid sequence is known Post-translational and process-related modifications are evaluated Relevant higher order structure, including disulphide linkages, are evaluated
Potency and Quantity The biological activity is measurable The quantity is measureable
Purity and Impurities The purity is quantifiable The impurities are identified, if technically feasible, and quantifiable

Characterization assays should be selected which give a maximum of information concerning the physicochemical and biological/biochemical properties of the drug substance. A much broader range of methods is available for characterization studies compared to specification assays [4]. Characterization assays can use sophistocated equipment run by an expert in a single laboratory (e.g. mass spectrometry) and research methods are appropriate. In contrast, specification assays must be robust and suitable for use by a large number of laboratories and analysts. Specification assays must be suitable for transfer between laboratories and sufficiently robust for inclusion in public standards (e.g. Pharmacopoeial monographs).

Product Specifications

Establishing appropriate product specifications is clearly one of the more important steps in defining a control strategy. The overall control strategy should accomplish the following: (i) ensure that the manufacturing process is consistent, (ii) ensure that product quality is maintained and (iii) ensure patient safety and efficacy. Product specifications should be established for those product attributes which affect quality, safety and efficacy and which are not adequately controlled by other elements of the control strategy (e.g. process validation, characterization, in-process controls, etc.). For example, many process impurities are better controlled by in-process testing (e.g. viral screening for mammalian cell culture) than by final product testing because the potential levels are greater at that point. Other process impurities, such as host cell proteins and source DNA, can be more effectively controlled by process validation studies which offer greater sensitivity and more meaningful data than final product testing.

Assays to be included in the product specifications should take into account information gained from drug substance and product characterization studies, stability studies, process development studies, and preclinical/clinical batch analysis [5]. Characterization studies provide knowledge of the important physicochemical and biological/biochemical properties of the drug substance, which will guide the selection of appropriate assays for routine product identity testing. Stability studies determine the degradation mechanisms and rates of formation of degradation products, which will aid in the selection of stability-indicating assays. Process development studies help to define which parameters can be controlled by process validation studies, without the need to perform final routine product analysis. The results of preclinical and clinical batch analyses can be used as a guide to select those product attributes which are most critical for controlling by final product testing (i.e. those attributes signficantly affect product quality, safety or efficacy and have been shown to vary significantly from lot to lot).

Numerical limits for specification assays should take into account preclinical and clinical batch analyses, manufacturing process capability (variability), assay method capability (variability) and product stability [5]. Appropriate statistical tools should be used to calculate meaningful specification values. Specifications at initial registration should be established primarily on the basis of results for preclinical and clinical trial batches, since this dataset is most directly linked to patient safety and efficacy requirements. After initial registration, as long-term consistency data are obtained for the full-scale manufacturing process, more rigid process control limits can be established, based on statistical evaluation of the full-scale manufacturing batches. Companies should establish internal release specifications which, taking into account product stability data as well as assay variability, ensure that the product will meet the regulatory specifications throughout its shelf-life.

Harmonizing Specifications

Harmonizing specifications on a global basis is a highly challenging and important exercise. Differences between the various regions which must be resolved include: (i) differing test requirements for a given test parameter such as sterility, (ii) different degrees of emphasis on process description/control and product specifications, (iii) requirements for repeating lot-release testing when importing into certain countries, and (iv) differing regulatory policies for reviewing process changes after product approval. For example, in the United States, many process changes can be implemented using the Annual Report as a mechanism for reporting the change to the FDA. No comparable mechanism exists in either Europe or Japan. As another example, Japan places great emphasis on product specifications and relatively less emphasis on the detailed process description and process controls, compared to the US and Europe. Some regions require re-analysis of a product for all specifications upon import; hence selection of appropriate specifications (i.e. limited to parameters which must be run on the final product to have adequate control) can have a significant impact on product cost. Each region, understandably, expects that manufacturers will use compendial tests listed in the relevant regional Pharmacopeia. Unfortunately significant differences exist in the test methods and specification limits between the various Pharmocopeias, thereby multiplying the assay requirements for global manufacturers and increasing product costs.

As the ICH process moves forward towards a specifications guideline it is important to recognize and deal with regional differences in an open and direct manner. In some cases, harmonization may be possible only if certain regulatory policies and/or laws are modified in one or more of the geographical regions. While changing laws can be a time consuming and complex process, in the long run such changes will prove worthwile since consumers, regulatory agencies and pharmaceutical manufacturers all stand to benefit from global harmonization.

References

1 International Conference on Harmonization. Harmonized Tripartite Guidelines: Stability testing of new drug substances and products. Step 5 Document (Effective January 1, 1998).

2 FDA Guidance concerning demonstration of comparability of human biological products, including therapeutic biotechnology-derived products, CBER, CDER, FDA, April, 1996.

3 Characterization of Biotechnology Pharmaceutical Products Conference, Sponsored by US FDA, December 11-13, 1995.

4 Davis GC et al: An evaluation of the methods and approaches used in the characterization of recombinant polypeptides for human pharmaceutical use, Lubiniecki AS, Vargo SA (eds): Regulatory Practive for Biopharmaceutical Production. New York, John Wiley and Son, 1994.

5 Davis GC et al: Rational approaches to specifications setting, an opportunity for harmonization. J Pharmaceutical Technology 1996;20(9):110-118.

Dr. G.C. Davis, Lilly Research Laboratories, Eli Lilly and Company, Lilly Corporate Center, Indianapolis, IN 46285, USA.

Brown F, Fernandez J (eds): Development of Specifications for Biotechnology Pharmaceutical Products. Dev Biol Stand. Basel, Karger, 1997, vol 91, p 55

WORKSHOP I

Environmental Control

Brown F, Fernandez J (eds): Development of Specifications for Biotechnology Pharmaceutical Products. Dev Biol Stand. Basel, Karger, 1997, vol 91, pp 57-58

Workshop Summary

J. Fernandez

The workshop on «Environmental Controls» included a definition of where environmental controls need to be applied; i.e. a closed system which can be validated. The group discussed how environmental validation evolves as the process develops and how those phases become continuous through such stages as sanitization, static conditions, dynamic conditions and equipment and validation of systems in routine operations. Defining the state of the system being environmentally monitored becomes the responsibility of the owner of the system, not the regulatory agencies. The history of the system and its indigenous organisms and their potential effects on product quality are the main drivers of understanding the quality of the environment and its changes during process evolution. For environmental controls to be effective, the monitoring programme needs to define the difference between bioburden and contamination at each stage of the process and this programme should include written reports of the justification for differentiating bioburden and contamination. Thorough sampling operations should reflect each of the manufacturing steps and also include normal operations in the manufacturing areas.

Workshop Slides

Environmental Control – Slide 1

- A closed process which has no external contact, precludes the escape of product, and prevents the entry of contamination could occur in a «Non-Microbiological Control» area, provided the process is validated to be a closed system. Closed system validation may be difficult but is a key point to its acceptance.

ENVIRONMENTAL CONTROL – SLIDE 2

- Environmental validation of a facility is a continuum which moves through several phases. Routine samples may be less than validation, but sampling never ends. Phases may include:
- Sanitization
- Static conditions
- Dynamic conditions and equipment
- Validation and routine operations

ENVIRONMENTAL CONTROL – SLIDE 3

- The FDA does not maintain a list of objectionable organisms. The level of organisms and the history of the system should be considered. Organisms become objectionable depending on their effect on product quality. *Pseudomonas* in a water system may or may not be objectionable depending on the production step and intended product usage, e.g. non-terminally sterilized topicals vs oral dosage units.

ENVIRONMENTAL CONTROL – SLIDE 4

- The monitoring program must define at each stage: When it is bioburden vs when it is contamination? Justification should be in writing
- Dynamic sampling occurs under normal operations. Normal operations can include maintenance, preparation work and general activity in the area, as well as specific manufacturing steps. Sampling should not avoid manufacturing steps.

Brown F, Fernandez J (eds): Development of Specifications for Biotechnology Pharmaceutical Products. Dev Biol Stand. Basel, Karger, 1997, vol 91, p 59

WORKSHOP II

Specifications: What is Necessary to Start Clinical Trials?

Brown F, Fernandez J (eds): Development of Specifications for Biotechnology Pharmaceutical Products. Dev Biol Stand. Basel, Karger, 1997, vol 91, pp 61-62

Workshop Summary

J. Fernandez

The workshop on «Specifications – What is Necessary to start Clinical Trials?» included discussions on product definition and process definition as well as the level of rigour required to initiate Phase I trials through Phase II and III trials. Confirmation of protein primary structure is highly desired for Phase I trials, but if development of the peptide map proves extremely challenging other forms of structural confirmation may be used. For early Phase I trials, removal of adventitious agents and contaminants is desirable, but process validation to confirm removal may not have been generated at the time trials start. Continuous data recording as the process evolves and contaminants are removed is an essential component of showing evolution of the process and removal of adventitious agents and contaminants. Previously validated processes may be used as a component of an evolving new process. Production and characterization of toxicology and Phase I materials need to be proved to be comparable; however a specification of «Report Result» is acceptable for non-safety related tests at Phase I and toxicology stages of development. At Phase I, DNA and host cell contaminants need not be quantified at low levels but evolution of the analytical procedures needs to show continued development of lower levels of quantification. Safety can be assured using general tests and methods (USP of similar.) Efforts should be made to identify low level process impurities. Recognizing that drug development is a continuum, the rigour of process validation and documentation required for initial toxicology and safety studies is not as demanding as those needed in subsequent Phase II and III trials.

Workshop Slides

Specifications: What is Necessary to Start Clinical Trials? – Slide 1

- Confirmation of primary structure is desirable, but if the peptide map is too complicated a combination of data can be used for structural conformation
- Adventitious agents and contaminant removal does not require a process validation for Phase I trials. Data should be generated from the beginning. Previously validated processes should be referenced.

Specifications: What is Necessary to Start Clinical Trials? – Slide 2

- The manufacturer needs to demonstrate that toxicology and Phase I materials are comparable.
- «Report Result» is an acceptable specification for non-safety related tests for toxicology and Phase I.
- Low level of quantification of DNA and host cell contaminants are not required for Phase I.
  General tests and methods can be used to assure safety.

Specifications: What is Necessary to Start Clinical Trials? – Slide 3

- The drug development process is a continuum. The rigor of process and documentation required for initial toxicology and safety studies are not as demanding as for subsequent Phase II and III trials.
- An effort needs to be made to try to identify low level impurities.

Brown F, Fernandez J (eds): Development of Specifications for Biotechnology Pharmaceutical Products. Dev Biol Stand. Basel, Karger, 1997, vol 91, p 63

SESSION III

Specifications Within the Manufacturing Operation

Moderator: *Florian Horaud, IABS*

Brown F, Fernandez J (eds): Development of Specifications for Biotechnology Pharmaceutical Products. Dev Biol Stand. Basel, Karger, 1997, vol 91, pp 65-72

Raw Material Considerations

A.S. Lubiniecki, P.J. Shadle

SmithKline Beecham Pharmaceuticals, King of Prussia PA, USA

Abstract: Raw materials (RM) testing and control strategy form a major part of the foundation of a well-characterized protein (WCP) or biopharmaceutical. Raw materials may be present in the final vial as excipients or may have product contact earlier in processing. Manufacturers of WCPs should use a scientific approach to set acceptance criteria and test methods for bulk raw materials which are not present in dosage forms in substantial amounts. These methods and standards should enable the procurement of specified RMs which enable the reliable preparation of a product which consistently meets its quality attributes. Manufacturers of WCPs must use available pharmacopoeial standards for excipients and bulk process RMs which cannot be substantially removed during purification or other processing steps. Manufacturers should support efforts to harmonize pharmacopoeial standards for excipients.

INTRODUCTION

Well-characterized biopharmaceuticals or biotechnological/biological products (WCPs) are defined by the International Conference on Harmonisation (ICH) as those whose structural features (including amino acid sequences, as well as physico-chemical, biochemical, biological and/or immunochemical properties) have been elucidated using a set of modern bioanalytical testing methods [1]. This is widely interpreted to mean that scientifically appropriate characterization studies have been performed, and that a subset of these methods have been selected for use as quality control (QC) and stability-indicating assays. In addition, it is usually assumed that a well-characterized biopharmaceutical is derived from a consistent process capable of validation, and that this process is operated in compliance with Good Manufacturing Practices (GMP). We suggest that there be an additional consideration – that the product is prepared by a process which employs appropriately chosen and controlled raw materials.

This modest proposal leads directly to several questions which must be answered in order to devise a rational approach to raw material control. The first is, «What quality standards are appropriate for RMs used to manufacture WCPs?». The second is, «Is the use of pharmacopoeial standards scientifically appropriate for bulk RMs used to prepare WCPs?». We will attempt to propose answers to these questions as we discuss the relevant issues.

In general, processes for WCP use living cells or organisms to generate the active ingredient. A large number of raw materials are used during bulk fermentation and purification, as will be illustrated later. Often complex or undefined RMs are used, some of which may also be of biological origin (e.g., hydrolysates, serum, organ extracts). Testing alone cannot assure the quality of the cells, raw materials, or products. The cell substrates derived from banked cells or source materials (derived from animal or human tissues or body fluids) are also raw materials and have special quality and safety concerns. In some cases, additional RMs may be used for chemical derivitization, final product excipients, and cleaning reagents. These cleaning agents may not contact the product stream directly, but nevertheless are controlled as RMs. Before we discuss standards, we would like to provide some background in raw material quality concepts.

Quality Concepts

The basic quality concepts appropriate to raw materials include assurance of the identity, purity, suitability, and traceability of all raw materials used in the process. Table 1 lists some of the principal tools available for providing this assurance.

Identity

The goal of RM identity testing is to verify that the article is what it is purported to be. The degree of difficulty in establishing identity is related to the number and complexity of the chemical entity(ies) in the RM. The typical methodology employed for chemically definable pure raw materials is physicochemical and provides unequivocal identification. For example, simple tests can establish the presence of sodium and chloride, and thus confirm the identity of a container of sodium chloride.

For complex or undefined raw materials, diverse testing methods are employed. The testing generally reflects both the nature of the substance(s) and the available analytical technology, and should be capable of discerning the substance of interest from likely substitutes from the same vendor. For example, immunochemical tests

Table 1: Raw material quality attributes.

Attributes	Tools to provide assurance
Identity	Testing, traceability, labels
Purity	Testing Inspection of item(s) Vendor certificate of analysis
Suitability for intended use	Process validation Vendor audit programme Performance testing if needed
Traceability	Vendor audit programme Vendor certification Certificates of analysis Contractual obligations under change control Labelling, control

with specific antisera can distinguish foetal serum from adult serum to confirm its identity, but cannot be considered to establish identity. Chromatographic profiles or fingerprints of amino acids or vitamins may be useful to identify some dehydrated media. Often, the level of information or assurance provided by identity testing is not complete. In a few cases, only performance testing (demonstration of successful preparation of a critical intermediate or material in a scaled down unit operation) can provide adequate assurance that the RM really is capable in the process, perhaps because its active ingredients are not well understood. For complex RMs, assurance of identity usually rests on an integrated combination of QC testing on incoming RMs, manufacturing controls (by the vendor), traceability, and confirmatory auditing of the vendor by QA on a periodic basis. No single element of this combination is sufficient, but collectively they provide strong evidence that the RMs used are of correct identity and quality. Manufacturers will always perform some type of identity testing on each lot to assure that a mix-up has not occurred at the vendor. It is not sensible for the vendor to perform the only tests of identity to assure that mix-up has not occurred.

Purity

The goal of purity testing is to determine that the RMs is not adulterated with significant levels of specific impurities or contaminants which would affect its intended use in the process. Most of the concepts described above for identity testing are also true for purity testing. In addition, specific concerns must be formulated in practical terms to permit appropriate specific testing. The extent of assurance needed reflects the criticality and potential impact of substitution at the point of use of the RMs. For example, there may be different purity requirements for sodium phosphate used as a formulation excipient, in purification buffers, in fermentation media, and in solutions used to clean process equipment.

Some RM are associated with safety concerns. For example, heavy metals can be a concern either for the process or for the product, and acceptance levels should be set for the RMs. Those RMs which are present in relatively low abundance may be of less concern than those which are more abundant, because the effects of these impurities are generally on a per-dose basis. For some RMs, specific biological concerns may require the use of appropriate testing for bioburden, endotoxin, viruses, mycoplasma and enzymes. During process development, it is important to establish concrete goals for purity testing which are attainable and which reflect the inherent limitations of the assays used. For example, it may be a laudable goal to state that the RM should be free of viruses, but this can never be proven in any laboratory due to the inherent limitations of sampling and virus assay methods [2]. An attainable goal is to state that the RM should be free of detectable viruses when tested according to a specific set of methods having established sensitivity. Thus, general goals should be defined in operational terms based on what can be demonstrated in the laboratory

Testing for purity or adventitious agents can be done by either the vendor or the WCP manufacturer, depending on the purpose of testing, the difficulty of testing, and the value of the test. Testing should be performed downstream of anticipated points of contamination. Virus screening assays for bovine viruses in animal serum can be done by most vendors in a credible manner, because there is no significant risk that finished serum stored frozen in intact containers will become contaminated with bovine viruses. There is no compelling reason to repeat the vendor's tests for adventitious bovine viruses. In contrast, testing for endotoxin or bioburden may be justified, since both of these contaminants can change upon storage of many RMs and the vendor's data may not accurately reflect the quality of the RM at its time of use.

Suitability

Another assurance which should be provided for a GMP raw material is its suitability for its intended use. Manufacturers should demonstrate that a specific RM can be used in a manufacturing process which can be validated to prepare intermediate or product which consistently exhibits specific quality attributes. This can be accomplished without specific studies by documenting the use of specific raw materials in the preparation of acceptable clinical supplies and qualification batches. For some complex (especially proteinaceous) RMs, it may also be necessary to consider the design of vendor processes for collection, processing, packaging, distribution, and storage of the substance and how these might affect use in the WCP process or the WCP itself. Performance testing may represent a useful approach for some complex RMs if simple standards for acceptability cannot be defined (e.g., for some chromatography resins, demonstration of a characteristic profile with process feedstock in a scaled down system). Again, an RM may have demonstrable quality for use as a cleaning agent yet fall short of the pharmacopoeial standards needed for direct parenteral use.

Traceability

Traceability is another quality attribute of RMs which is required by the GMPs. The manufacturer of a WCP should be able to trace and document the origin of all RMs back to a specific process used by the ultimate source (primary manufacturer) of the RMs. Traceability helps to assure not only the identity and purity of the RMs, but may define the likely contaminants as well. For complex undefined raw materials, the low power of analytical tools to detect contaminants of concern places additional emphasis on traceability to assure quality. For animal-derived RMs, concern over putative contamination by viruses and prions has increased the importance of traceability and certification of origin as a control tool. A functional vendor audit programme must be developed early in product development and may add more safety than direct QC testing. Vendor certification combines these elements with assay cross-validation, change controls, and other contract use procedures to provide maximal assurance of RMs quality [3].

Standards

With the above background, we can discuss the subject of quality standards for RMS. What level is appropriate? None? Appropriate science? Pharmacopoeial standards throughout the process? Case-by-case? Some combination of these? For active ingredients and for excipients, the requirement for compliance with pharmacopoeial standards (when available) is clear, obvious, and inarguable. But what about RMs used in bulk processing, or cleaning? Should they meet the same standards as parenteral drugs? Does the added cost of testing, controlling, and purchasing these add to patient safety or quality assurance? This is not a theoretical question, for some national control authorities and application review bodies have required that all RMs (including those used in fermentation) comply with local pharmacopoeial standards when monographs exist for an RM. Until full harmonization of the major Pharmacopoeiae has been accomplished, this approach poses significant practical difficulties. In contrast, where a monograph does not exist, the manufacturer formulates and justifies standards that seem to be scientifically appropriate, and demonstrates the appropriateness of these standards by collecting process consistency data.

The most pressing concern for a multinational WCP manufacturer is determination of which Pharmacopoeia to follow. The major Pharmacopoeiae (USP, EP, JP, for example) currently differ in both individual specifications of many common raw materials, but also in methods such as RM sampling methods, rules on number of samples, sample compounding, and sample handling. To meet more than one monograph then requires double sampling and testing for the same quality attribute. Water for injection (WFI) is defined by different analytical methods and standards, including pH method; hence it is conceivable that a sample could pass by one test and fail by another. How the manufacturer then determines whether the water is WFI or not is unclear. Methods and calibration standards for many RMs also differ among the major Pharmacopoeiae. Table 2 shows the coverage of pharmacopoeial monographs for the RM used to produce a typical WCB. About half the RMs had monographs, while all formulation excipients had monographs. Most purification RMs except chromatography resins had monographs. Thus, it appears that the prevalence of monographs increases as the process moves closer to the final product. This is not surprising from the point of view of a biologist (that cultivated cells would have somewhat different growth requirements than could be found in the pages of Pharmacopoeiae) or from that of a formulation scientist (who will always use pharmacopoeial ingredients if available).

Comparing the monographs of two Pharmacopoeiae for these RMs in Table 3, nearly half (26) have no monograph in either volume. For these RMs the manufacturer sets the standards and verifies appropriateness by process performance and

Table 2: Availability of pharmacopoeial monographs for raw materials for a typical WCP.

Process area	**# RM In pharmacopoeia**	**# RM Not in pharmacopoeia**
Cell banking	6	5
Basal medium	1	(70 components)
Fermentation	25	17
Purification	15	4
Formulation	7	0

Table 3: Analysis of pharmacopoeial monograph requirements.

Availability of monograph	**#**	**Outcome**
Not in either pharmacopoeia	26	Manufacturer sets standard
In both pharmacopoeiae	28	Several, including:
	15	Accept highest spec; or, Method/standard crossover
	12	Significant resources to prove RMs meets both; may need to specify a new RM quality
	1	Mutually exclusive standards; Different RMs required for compliance in each area

vendor auditing. For 28 RMs, both Pharmacopoeiae had monographs. Of these 28, 15 RMs had monographs which were sufficiently similar to accommodate by accepting the higher standard of each pharmacopoeia. For 12 of the 28, the monographs required significant methods and standard cross-over to demonstrate that a single RM met both monograph requirements (usually duplicative testing to different requirements, or changing to a new RM source). For one RM, the standards were mutually exclusive; compliance with both Pharmacopoeiae would require that the WCP be made by two processes, each with a different RM for each of these affected markets unless an exception could be justified. This action would have significant effects on manufacturing costs and on production/inventory planning.

By design, pharmacopoeial standards are intended for application to active ingredients and excipients found in drug products. Another interesting question is whether pharmacopoeial standards are a guarantee of suitability for intended use. For a few applications such as for serum-free cell culture, pharmacopoeially-specified purity levels may be inadequate to permit cell growth because no standard is set for this application, and requirements will vary among cell types. Different products and processes may well require different purity levels for the same RM. Some pharmacopoeially-specified characteristics may be irrelevant for certain applications, and represent overspecification or inappropriate regulation for others (e.g., endotoxin specifications for RMs used in *E. coli* fermentations, where the organism is 4% endotoxin by weight). Certainly, it will be difficult for the Pharmacopoeiae to anticipate all possible uses for a bulk RM in a monograph. Thousands of monographs would need to be created for RMs not currently present in the Pharmacopoeiae. Given the length of the revision cycles currently, this project would clearly take a long time and might lengthen the time required to revise other monographs due to congestion of the system.

Looking at other types of pharmaceutical products, pharmacopoeial standards are generally not applied to chemicals used to prepare bulk drug substances. Further, it is not clear what types of problems are prevented or minimized by advocating the application of pharmacopoeial standards to bulk processing, since no adverse events or known problems associated with WCPs can be linked to RMs standards. Finally, WCP manufacturers currently establish RM specifications and standards for a significant portion of their RMs (in our example, about 50%) which are not listed in any monograph. Clearly, there is no practical impediment to manufacturers establishing science-based specifications and standards for 100% of bulk RMs used to make WCPs.

Proposal

We propose that WCP manufacturers develop standards and specifications for their RMs based upon a rational science-based approach considering the nature of each RM and its intended use(s), and what can be achieved realistically in the laboratory using available validated or qualified assays. Final product excipients (and active ingredients where such monographs exist) require pharmacopoeial specifications and standards when available and require available appropriate scientifically-based standards when monographs are not available. Pharmacopoeiae should harmonize methods and standards for monograph items as soon as possible to facilitate international agreement on acceptable standards. Bulk RMs should have standards which are based on their process «fate», which we define in the next paragraph. Finally, on a case-by-case basis, specific controls for specific concerns may be appropriate (e.g., viruses and prions for animal-derived RMs, bioburden and endotoxin levels for cell culture RMs).

For RMs used in fermentation or purification prior to formulation, the standards should reflect the «fate» of the substance. In other words, if the RM is substantially removed from the product or process stream by further processing, or is substantially transformed into another chemical entity by chemical, enzymatic or metabolic processes, then the standard applied to that RM should be scientifically appropriate and suitable for the intended use of the RM. This may differ from a pharmacopoeial monograph of the same RM when used as an excipient. This definition of «fate» now begs an explanation of «substantial», which we would propose to mean removal or transformation to levels which are: (i) expected to be non-pharmacologically active based on literature results or preclinical studies, and (ii) not more than 1% of their original concentration with respect to the concentration of the active ingredient at harvest or point of RM use (whichever is greater). Our choice of a 1% level is arbitrary, but there is no rational scientific basis for any other number in general, and this 1% value would seem to be easily defined, measured, and used consistently. If a bulk process RM was substantially present in the final product, then pharmacopoeial standards should be used for the RM where available. Naturally, pharmacopoeial methods may be used for convenience as desired. It may not be beneficial to have three different specifications for NaCl, in general, if it is used early and late in a process. However, such decisions should be made based on scientific and practical grounds by the manufacturer who is familiar with the use of the RM and the data for the process/product, rather than by a group of pharmacopoeial experts unfamiliar with a given use and the supporting data, and who really cannot be expected to anticipate all such uses.

Summary

Manufacturers of WCPs should use a scientific approach to determine the standards for bulk raw materials which are not present in dosage forms in substantial amounts. These methods and standards, coupled with RM controls on traceability and a strong vendor audit programme, should enable the procurement of specified RMs which enable the reliable preparation of product which consistently meets its quality attributes. Manufacturers of WCPs must use available pharmacopoeial standards for excipients and bulk process RMs which cannot be substantially removed during purification or other processing steps. Manufacturers should support efforts to harmonize pharmacopoeial standards for excipients, and to cross-validate methods and standards which are unique to each pharmacopoeia.

Acknowledgement

We gratefully acknowledge the help of our colleague Jan Wood in preparing this manuscript.

References

1 International Conference on Harmonisation Expert Working Group on the Quality of Biotechnology Products. Annex to the Tripartite ICH Guideline for the Stability Testing of New Drug Substances and Products. Step 4 Document. 1995.

2 Lubiniecki AS: Current issues in virus assays and viral clearance evaluation. Dev Biol Stand (in press).

3 Pharmaceutical Manufacturers Association. Guidelines for Vendor Certification. Washington DC. 1989.

Dr. A.S. Lubiniecki, SmithKline Beecham Pharmaceuticals, King of Prussia, PA 19406, USA.

Brown F, Fernandez J (eds): Development of Specifications for Biotechnology Pharmaceutical Products. Dev Biol Stand. Basel, Karger, 1997, vol 91, pp 73-78

In-Process Testing and Limits

I. Sjöholm

Department of Pharmacy, Medical Products Agency, Uppsala, Sweden

Key words: Biotechnology, in-process controls, specifications, quality assurance.

In-process controls are important instruments to ensure qood quality in any pharmaceutical manufacture, in particular the production of biotechnology pharmaceutical products. This paper deals with the role of in-process controls in biotechnology manufacture, the rationale for performing these tests and – to some extent – how to design and use them. The views presented are in accordance with the current general view in Europe and particularly at the Medical Products Agency in Sweden. However, it is important to emphasize that each specific problem has to be handled case by case. The views presented here, are therefore not necessarily the final views of the Committee for Proprietary Medicinal Products (CPMP) on a specific issue within the European Community.

Looking back 10-15 years, one can conclude that the quality of drugs and the consistency in the production of an individual drug have been improved dramatically. The regulatory agencies and the manufacturers have arrived at the common and correct point of view that Good Manufacturing Practice (GMP) and acceptable systems for Quality Assurance (QA) – are necessary to guarantee, as far as possible, the correct quality of a drug. Everyone now realizes that quality has to **be built into** a product. Quality cannot be guaranteed by controlling only the final product in a batch release procedure. The in-process controls will therefore be an essential part of any QA system.

It is obvious today that the introduction of biotechnological procedures for the production of insulin and somatropin in the early 1980s meant a break-through for the understanding and acceptance of the concept that consistent manufacturing conditions, together with careful and appropriate in-process controls with validated methods, are key factors in the quality control of biotechnology drugs and also for drugs in general. For instance, the need for in-process controls of plasmid stability, fermentation procedure, freedom from DNA and host/vector cell proteins and a close follow-up of the down-stream purification of the protein are well identified and accepted by everyone as important aspects in the production of biotechnology products. It can even be said that in-process testing is the most important part of quality control.

Guidelines

The regulatory network for medicinal products in the European Community (EU) is now well developed. For in-process testing, the most relevant guideline is called Notice to Applicants containing instructions for what shall be included in the files when a licensing application is made in any member state of the EU. This guideline is currently being revised and it will doubtless undergo regular revision. Of particular interest is the annex Biotech Headings, Part II, (III/3153/93-EN). This annex gives advice on when in-process controls and the limits applied shall be specified in the documentation, for example, the controls performed in the manufacturing process (section Part II B. 2. Method of Preparation. Manufacturing process) is mentioned as an important part, where in-process testing is relevant. The same is true for several sections in Part II C. Production and Control of Starting Materials. As a principal point of the guideline, it is stated that «All procedures need to be validated and the results of the validation studies must be provided».

The Need for In-Process Testing

The introduction of new, improved production methods for biologicals in general, despite the development of modern analytical methods, has made it quite clear that adequate tests on some properties of the active compounds or on important contaminants cannot always be performed on the final product. This problem arises, for example, when one has to modify an active substance during the formulation in such a way that it is no longer possible to perform any precise assay on the final product. Such a situation is true for example for detoxified vaccines or allergens precipitated with alum. A similar problem is seen with products to which auxiliary substances are added, for example albumin in interferon preparations, to improve the stability of the active substance or minimize unspecific adsorption to glass walls or tubes. In these situations, a selective assay has to be performed in-process on the bulk just before the modification of the compound or the addition of the auxiliary substances.

In the pharmacopoeias, the increased need to perform in-process testing has also been observed. Tradition maintains that the testing and formal release of a batch of a preparation shall and must be done by testing on the final substance or formulation. This is still true for most of the synthetically produced raw materials and can be done adequately with precision and methods that even allow the identification of the place of its production or the manufacturer, by making a finger-print of the impurities. However, for many biologicals the pharmacopoeias today have included sections which specify requirements related to the production processes. These sections often describe tests that have to be validated and performed in-process and, in addition, are quite complex sometimes and not suitable for routine testing in a release procedure.

Control of the Manufacturing Procedure

The consistency of production is one of the key points for achieving the correct quality of the product. Some factors to be tested during manufacture are summarized in Table 1. It is well understood that even apparently minor variations of the parameters used to control a process, for example, stirring, buffer concentration or temperature during a fermentation procedure, pumping speed during a chromatographic separation, can profoundly change the yield or purity of the desired protein. Changed performance of a separate step in the manufacturing chain can also subsequently affect the next steps in the chain and as a final result also change the purity or properties of the protein in such a way that the product cannot be approved. Under such circumstances it is obvious that an early test of the performance of the critical steps in the manufacturing chain in order to collect information on the yield, purity or degree of contamination or whatever is to be controlled, will save considerable money and time.

The limits to be used in the testing can be based on the results of the validation of the manufacture. Admittedly, the information may be limited to a small number of batches, when the licensing application is filed, but should be collected from full-scale runs in the plant, where the commercial production will take place, and revised when necessary. A continuous follow-up of the test results is advisable, which can give information about any negative – or positive – trends in the manufacture to allow action to be taken.

In-Process Testings on the Active Protein

Generally, when one talks about in-process controls, the focus is on the active protein and what is happening to it during manufacture. Indeed, a main reason for performing the in-process control is to ascertain that the correct product is obtained in a yield and with the quality anticipated from the validation studies.

For biotechnology products, a great deal of interest was focussed initially on methods used to prove that the plasmids transfecting *E. coli* were stable. This required an in-process control by restriction enzyme maps. In several cases, this test was proved to be a very useful test to show that an appropriate vector construction had been chosen to allow a stable and correct production of the protein. With incorporated/fused gene constructs in mammalian cell cultures, the in-process controls

Table 1: Tests to be performed in-process during the manufacturing process.

In-process control of biotech proteins
* Fermentation
* Yield
* Plasmid stability
* Absence of live cells in harvest
* Identity
* Purity
* Potency, content

are now more focussed on the control of the protein yields and structural analyses of the proteins. The in-process controls have to show that after proper validation, the parameters established for the fermentation are followed and give the anticipated yield of intact protein. In mammalian and yeast cells, the yield is still considered to be an important indication of a correct fermentation giving a consistent carbohydrate conjugation, both as far as the site for the conjugation and the structure of the carbohydrate side chains are concerned.

An important test after all fermentations, which is often forgotten, is to check that no living cells contaminate the raw product, that is that the killing, disintegration or separation of the cells from the harvest are complete. The stability and purity of the active component during the purification is often highly dependent on the absence of any proteolytic enzymes in the first cell extract.

Furthermore, it is imperative that the chromatographic separations or precipitations by salt or organic solvents during the down-stream purification give the anticipated results. A selective assay is needed for the active protein, so that a calculation of the specific content is possible. It is presumed, of course, that the validation of the manufacture during the development has shown that a protein with full biological activity is obtained, as ascertained with a set of bioassays in animals, cells or with other suitable, biological methods. However, in routine testing and for the determination of the content of active protein in the final formulation, more precise physico-chemical methods have to be chosen, once such methods are available, instead of the generally more laborious, less selective, less reproducible and more expensive bioassays. As pointed out earlier, it may be necessary to perform the final, selective assay as an in-process control before the final formulation and filling, if these procedures make an assay impossible. It may also be advisable to make the final assay of the content immediately before the final dilution and dispensing, if this step is simple. The correct dilution can then be checked by a simple UV-control.

In-Process Testing of Contaminants

When the first products prepared via recombinant techniques were licensed, the main focus was on potential contaminants, some of which are listed in Table 2. Setting limits for DNA, foreign protein or virus contaminations are always a nuisance for regulatory agencies as well as for manufacturers. These contaminants have caused and are still engendering considerable work and debate, because there is no concrete information on which to base sensible limits. In general, as a simplification, one can say that the approved contamination limits for host cell and vector DNA and foreign proteins, in biotechnology products, do not present serious safety risks at present. However, virus contamination – as is well known – may give rise to problems, despite all the guidelines, and precautions such as virus inactivation procedures. New viruses or variants which have not been foreseen and which may transmit infections to patients, will always be detected and identified. These are consequently reasons for being cautious and for using a battery of tests and precautions according to the published guidelines so as to prevent all possible contaminations.

The most important step in achieving the correct quality of the product and to decrease safety risks is adequate validation of manufacture. It is important to identify the specific steps in the down-stream purification which remove or inactivate

Table 2: Contaminants which should be tested for in-process control.

Contaminants
* DNA, foreign proteins
* Virus
* Bioburden
* Pyrogens, endotoxins
* Reagents, organic solvents
* Column materials

viruses or functional DNA. The in-process testing is then essential to ensure that these critical steps are effective.

The control of foreign proteins can be performed in different ways – normally by immunometrical methods based on immune sera prepared from mock preparations. These tests are certainly not better than the antibodies used. It is therefore important to stress that in the validation of the method it is more important to perform it correctly, to show that the analysis is detecting relevant proteins, rather than to specify a low limit, which may have no meaning. Indeed, no safety problems originating from contaminating foreign proteins have been identified so far. However, one should be aware of the risk that foreign proteins may form complexes with the active component. Such complexes can obviously severely impair the purification procedure and – if not removed – may increase the immunogenicity of the product when used in patients.

DNA contamination from bacteria seems not to be of any great significance, as such DNA is normally not functional in humans. However, DNA from mammalian cells is a controversial subject and different, acceptable limits – 10 or 100 pg per dose – have been specified in different guidelines. The main point is to ascertain that the DNA is not functional, which essentially is a matter of the size of the DNA. Quite sensitive tests have been developed and should be used during the validation to ascertain that high molecular weight DNA is removed. Another issue is how the validation should be done, e.g. how data obtained after spiking with labelled DNA shall be interpreted. One should be cautious in adding the removal factors obtained for the different purification steps, if each step has been spiked with the same labelled DNA preparation. It is certainly essential that the in-process test is used early after the critical step in the down-stream purification, to ascertain that the removal/inactivation has been effective.

The tests for foreign protein, DNA and virus discussed so far can be regarded as so called GMP-related tests, that is once established to be safe, the limits are generally not very important, but compliance gives an indication that the manufacturing and purification procedures have been carried out correctly.

The purpose of the in-process testing on the microbial burden is different. Indeed, the aim of such tests is both to prevent microbial contamination which would degrade the protein during the purification procedure and to ascertain that the microbial load on the filters used to ensure aseptic conditions is not too high. The limit normally applied in Sweden is NMT 100 total viable counts per ml.

Hopefully, pyrogen testing can be avoided after appropriate validation of the manufacturing process and be replaced by LAL testing. Normally, the limits for endotoxins do not cause any concern for biotechnology drugs, and it should be possible to comply with a maximal limit of 5 EU per kg in a dose, at least for proteins which are not given in very large doses.

Other in-process tests, which are important to apply during manufacture, are those which are designed to detect reagents, organic solvents, antibiotics, antibodies or haptens used during purification. Affinity chromatography columns with covalently bound ligands always leak some of the protein into the eluate and such contaminants as well as organic solvents have to be limited. Also in these cases, the best step to test for these contaminants is as close as possible to the process, which is designed specifically to take care of the specific contaminant.

As to the specification of specific limits in the in-process testing, it is not normally possible to specify explicitly the maximal, acceptable level beforehand. Such limits have to be based on what is practically achievable, judged from three to five separate batches or runs. Admittedly, this principle will mean that the limits specified in the original licensing application are tentative and may have to be modified, when more experience has been gained from large-scale production.

It is important to emphasize that it is the whole set of analyses used during the manufacture, that is in-process controls and tests on the final formulation, which will determine whether the product has the correct and consistent quality and content. In-process testing is preferred in most situations at the sacrifice of final product testing. The set of tests used on the final product must be such that a good set of data representing 0-time are obtained to allow a close follow-up of the stability of the product during the entire shelf-life, if any unexpected reports from the market about adverse reactions or quality-related problems are forthcoming.

Conclusion

It is fair to conclude that the results from the in-process controls – based on proper validation studies and sound specifications – give the manufacturer **and** the control authorities the best instrument to assess the quality of a separate batch of a biotechnologically derived protein.

Prof. I. Sjöholm, Tallmovägen 14, S-756 45 Uppsala, Sweden.

Brown F, Fernandez J (eds): Development of Specifications for Biotechnology Pharmaceutical Products. Dev Biol Stand. Basel, Karger, 1997, vol 91, pp 79-88

The Use of Bioassays for the Characterisation and Control of Biological Therapeutic Products Produced by Biotechnology

R. Thorpe, M. Wadhwa, A. Mire-Sluis

Division of Immunobiology, National Institute for Biological Standards and Control, Potters Bar, Herts., U.K.

Abstract: The primary value of bioassays is that they alone directly assess biological activity of bioactive substances and products. Appropriately designed bioassays reflect the fundamental aspects of the biological activity of a bioactive molecule, including ligand–receptor binding, signal transduction processes (often poorly understood) and the final observed biological effects. Biological assays therefore complement physicochemical and biochemical procedures which normally only assess precise molecular structural features of complex molecules produced by biotechnology. Bioassays provide valuable information concerning the potency of biological products. This is essential for evaluating batch-to-batch consistency and stability. Bioassay data are crucial at all stages in the development of biological products, from early research work to final quality control of finished products. However, the type and design of bioassays may differ according to the information required and its intended use. The assays may or may not directly relate to the clinical use of the product. Bioassays can be difficult to perform and be time consuming, although this often reflects bad assay choice and/or design. Correct analysis of the assay results is essential if valid data are to be obtained. Standardisation, using correctly calibrated primary and secondary standards, is essential. In vivo bioassays are normally more unreliable than in vitro procedures, but in some cases the latter are either not available or do not address important biological characteristics of a product. Bioassays must be validated for their intended purpose and for the types of samples to be measured. Appropriate statistical analysis should be used to derive the significance and specifications of results. This needs to address both variability in samples and assay performance. Specifications (limits) for product acceptability need to be derived from real data using several batches of product.

INTRODUCTION

Unlike chemical drugs, biological products cannot be completely characterised by physicochemical methods alone. This necessitates carrying out one or more bioassays at various stages of product development and quality control to provide an indication of the biological activity of the substance. It is essential that such assays are correctly selected, carried out and analysed if valid and accurate potency estimates are to be obtained [1].

Bioassays – Essential Characteristics

A bioassay is a procedure for measuring the potency of a biological substance. It can be carried out in vivo or in vitro. A functional response is measured and analysed to provide a statistically qualified potency estimate. Bioassays involve measuring the response of a living system to a biological substance, which distinguishes them in a unique and important way from physicochemical methods and procedures which simply involve binding, e.g. immuno-assays and receptor binding assays [2]. This affects the results obtained and their significance [3]. Appropriately designed bioassays assess the fundamental activity-related aspects of biological molecules, including ligand-receptor binding, signal transduction processes and the final observed biological effects. The signal transduction mechanisms are often complex and poorly or incompletely understood and it is only by the use of bioassays that the capacity of a biological molecule to «trigger» these pathways can be evaluated. It is this important aspect of bioassays that mainly distinguishes them from all other methods, including receptor binding assays.

In Vivo and in Vitro Bioassays

In vivo (whole animal) bioassays can provide valuable information concerning the potency of a wide range of biological products. However, such procedures are often relatively imprecise, time consuming, expensive and tedious to carry out. The use of the fairly large number of animals necessary to provide an acceptable valid potency value can also be problematical and may have ethical implications. In most cases it is better, or at least preferable, to adopt an in vitro bioassay, although there are some apparent exceptions.

In vitro biological assays can be carried out using isolated organs or tissues, primary cultures of cells or continuously growing cell lines. Whole organ assays suffer from most of the problems associated with whole animal assays and are best avoided if possible. Assays involving the use of primary cultures of cells can be useful, but inconsistencies due to variation between different cell preparations can be problematical. Methods based on continuously growing, factor-dependent, cloned cell lines usually offer the greater precision, and accuracy as well as being the most economical and easy to use and analyse [2, 4, 5].

The Scope of Bioassays

Bioassays are used in all disciplines of biology for the determination of biological potency and related parameters. Although they are well established for use with «conventional» biological therapeutic products like vaccines, coagulation factors and hormones, their use with biotechnology-derived products such as cytokines is less well documented. However, as indicated above, bioassays provide unique information concerning biological potency. If these data are required, then establishing and using an appropriate bioassay is the only available valid approach, and is just as necessary for a very pure rDNA expressed protein as for less pure materials derived using «conventional» manufacturing methods.

Types of Bioassays

Theoretically, any biological property of a molecule can be used as the basis of a bioassay if it can be linked to a quantitative measurable endpoint. However, some biological effects are obviously more easily and accurately measured than others. In practice, a wide range of systems has been developed, which can be chosen according to the desired experimental format, available equipment and personal choice. Even for a relatively recently discovered class of biologicals like the cytokines, a fairly wide range of alternative bioassays is available (see Table 1).

Bioassay procedures and protocols for use with biotechnology products are best selected to produce low «background» responses, with steep reproducible dose response curves for the relevant product (see Fig. 1). Continuously growing factor-dependent lines often show these desirable characteristics. Factor-independent lines and primary cultures of cells often produce high backgrounds which reduce the significance of the signal and can also be variable in response to the product.

Bioassays – Expression of Results and Standardization

Analysis and expression of results obtained using bioassays obviously depend to some extent on the methods adopted. However, the approach of comparing dose-response curves obtained with batches of product and preparations of known analyte content is particularly valuable and informative. This involves assaying various dilutions of samples within the same bioassay – samples containing the same concentration of bioactive analyte will produce overlaying curves (see Fig. 1), whereas preparations containing different concentrations of biologically active product will produce displaced curves along the dilution axis [1]. This displacement is proportional to the difference in analyte concentration of the samples undergoing analysis (see Fig. 2).

Table 1: Types of bioassays available for cytokines.

1.	Proliferation assays
2.	Cytotoxicity/cytostasis/antiproliferative assays
3.	Colony formation assays
4.	Antiviral assays
5.	Chemotaxis/chemokinesis assays
6.	Induction of secondary effects e.g. HLA class I and/or II induction Secretion of other cytokines/molecules Synthesis of prostaglandins Respiratory «burst» Upregulation of expression of adhesion molecules

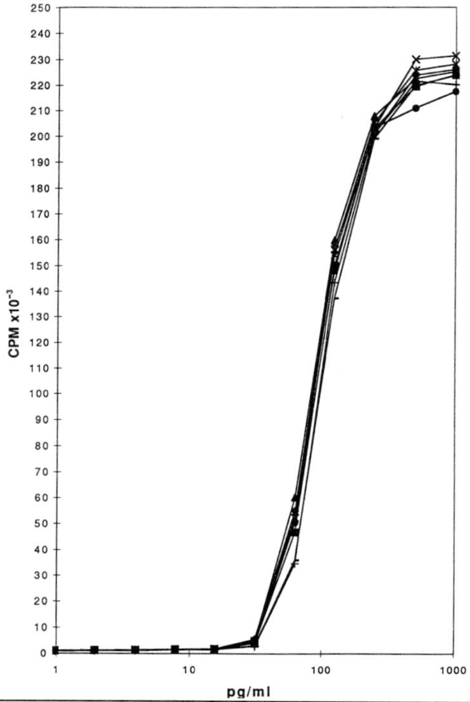

Fig. 1: Dose-response curves obtained using a cell line-based bioassay for human IL-15. All preparations assayed (which were formulated differently) produce essentially overlapping dose response curves showing that they have identical potency. Cell proliferation (of CTLL-2 cells) in response to rDNA-produced IL-15 was assessed by tritiated thymidine incorporation into nucleic acid [4, 5].

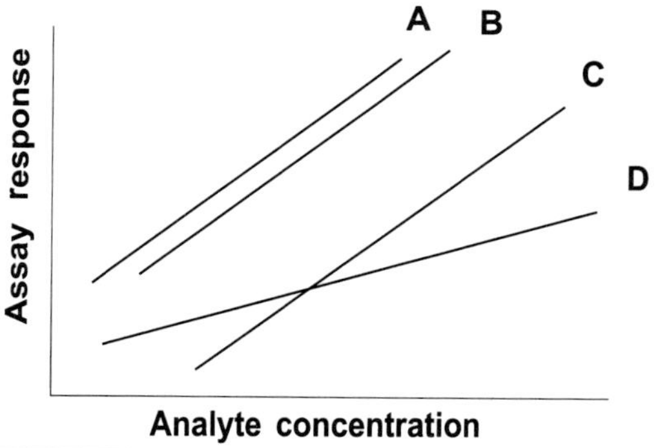

Fig. 2: Comparison of the linear portions of dose-response curves obtained using bioassays. Preparations A, B and C produce parallel lines, showing that the assay is responding similarly to the stimulatory activity in the preparations. The difference in displacement of the curves along the analyte concentration axis is due to different potencies of the three preparations; this is proportional to the analyte concentration in the samples. Preparation D is clearly not parallel to the other curves, which indicates that the bioassay is responding differently to the stimulatory material. It is not valid to compare the potency of preparation D with preparations A, B and C.

The dose response curves obtained with samples to be compared must be parallel; if this is not the case then the bioassay is responding differently to substances present in the samples being assayed (Fig. 2). Estimation of parallelism of curves obviously requires analysis of several dilutions of each sample and a comparison of the generated dose-response curves. Single «point» estimates, i.e. assays carried out using a single dilution of sample are not only unlikely to generate sufficient data to produce a valid estimate of potency, but also do not allow an assessment of parallelism of response and hence validity of the comparability of the samples under analysis.

Almost all bioassays are comparative assays. Their calibration therefore requires the use of a standard (or reference preparation) of known potency. «Calibration» of bioassays in terms of parameters such as 50% maximum proliferation etc., is not recommended as this effectively ignores the potency of the standard. A range of such potency standards are available which are suitable for the calibration of bioassays which can be used for products produced by biotechnology. Many of these have been produced under the auspices of the World Health Organization, e.g. for cytokines (see Table 2). Such standards are calibrated in units or International Units which are often arbitrarily defined and relate to the ampoule content of analyte [6]. The standards are used to calibrate bioassays either directly or for calibration of «secondary» or working standards. The use of a single primary standard worldwide facilitates the comparability of assay results. Standards must be included in each bioassay run and the potency of unknown samples can be calculated by comparison of the displacement of their dose-response curves from that generated using the standard (see Fig. 2). The dose-response curves of the standard and unknowns must be parallel if the comparison is to be valid. In some cases, mathematical manipulation can be used to derive parallel responses [7].

Table 2: WHO International Standards and Reference Reagents available from NIBSC for cytokines.

Preparation	Code
INTERLEUKIN 1 ALPHA rDNA (International Standard)	86/632
INTERLEUKIN 1 BETA rDNA (International Standard)	86/680
INTERLEUKIN 2 Cell line derived (International Standard)	86/504
INTERLEUKIN 2 rDNA	86/564
INTERLEUKIN 3 rDNA (International Standard)	91/510
INTERLEUKIN 4 rDNA (International Standard)	88/656
INTERLEUKIN 5 rDNA	90/586
INTERLEUKIN 6 rDNA (International Standard)	89/548
INTERLEUKIN 7 rDNA	90/530
INTERLEUKIN 8 rDNA	89/520
INTERLEUKIN 9 rDNA	91/678

Table 2: WHO International Standards and Reference Reagents available from NIBSC for cytokines.

INTERLEUKIN 10 rDNA	93/722
INTERLEUKIN 11 rDNA	92/788
INTERLEUKIN 12 rDNA	95/544
INTERLEUKIN 13 rDNA	94/622
INTERLEUKIN 15 rDNA	95/554
M–CSF rDNA (International Standard)	89/512
G–CSF rDNA (International Standard)	88/502
GM–CSF rDNA (International Standard)	88/646
LEUKEMIA INHIBITORY FACTOR rDNA	93/562
ONCOSTATIN M rDNA	93/564
STEM CELL FACTOR/MGF rDNA	91/682
BONE MORPHOGENETIC PROTEIN–2 rDNA	93/574
RANTES rDNA	92/520
GRO ALPHA rDNA	92/722
MCP–1 rDNA	92/794
MIP–1 ALPHA rDNA	92/518
IFN ALPHA LEUKOCYTE (International Standard)	69/19
IFN ALPHA 2B rDNA (International Standard)	82/576
IFN ALPHA D rDNA (International Standard)	83/514
TGF BETA 1 rDNA	89/514
TGF BETA 1 (NAT BOVINE)	89/516
TGF BETA 2 rDNA	90/696
TNF ALPHA rDNA (International Standard)	87/650
TNF BETA rDNA	87/640

Bioassay-derived potency data must be accompanied by a valid statement of the uncertainty associated with the measurement. Statistical procedures have been devised for the calculation of the degree of error associated with such determinations. This is normally expressed as a statement of the statistical confidence limits of the potency estimate. Bioassays must be carefully designed to minimize and allow an assessment of the errors associated with the derived potency value.

The Use of Bioassays at Various Stages of Product Development

In several cases, molecules which have become successful biological products were originally identified using bioassays or biological systems, e.g. the colony-stimulating factors, IL-2. It is therefore not surprising that such assays can provide useful, unique information at all stages of product development.

Biological assays of various types can be used at the very early stages of product identification and discovery. These can provide information of the types of activities shown by the potential product. During product development, bioassays are invaluable for investigations aimed at characterising the biological activities of the product and for stability, dosing and formulation studies. After product development, the assays are used to show batch-to-batch consistency and product (final form) stability under the proposed storage conditions as well as defining the biological potency of the product.

Different bioassays may be needed for different purposes at different stages of product development. Biological assays used early in product identification and characterisation may need to be designed to maximise biological relevance and information content and could be less precise, accurate and rugged as well as being different from assays used later in product development to determine potency, stability and batch-to-batch consistency.

Although some bioassays can provide information relevant to the clinical potential and use of the product, clinical efficacy and safety must obviously be directly assessed in clinical trials in humans. It is therefore not necessary that bioassays used for product control reflect the clinical indication(s) for which the product is used. It is usually best to select a good bioassay, irrespective of clinical relevance, especially for determining biological potency, stability and batch-to-batch consistency.

Measurement of the Biological Potency of Biological Products Produced by Biotechnology

Measuring the biological potency of batches of products is obviously important for the release specification of biological products. For this, a validated bioassay is used and analysed to provide a mean potency estimate and also fiducial limits of error for the batch. The criteria which must be satisfied by the mean potency estimate for each batch of product are often presented as a statement of estimated potency (with acceptance limits) and the acceptable fiducial limits of error around these. This approach is well established for many «conventional» biologicals, e.g. hormones, and is particularly appropriate for use for biological products produced by biotechnology. An example for a particular product could be that the estimated potency is not less than 80%, and not more than 125% of the stated potency (see below) and the fiducial limits of error are not less than 64% and not more than 156% of the stated potency. The format of such statements can be found in specification protocols for a wide range of products and in pharmacopoeial monographs. The implication of the above is that the mean potency estimate must lie within specified limits or the batch is unacceptable. These limits should be derived from the results of the mean estimate of potency for several different batches. This specification limits the batch-to-batch variation found with the potency of the product. The fiducial limits of error are derived from the individual potency estimates from which the mean estimates are obtained; they are a limit to the assay variability. For the purposes of bioassay analysis, fiducial limits can be considered numerically identical to confidence limits.

Specifications for limits are best based on real bioassay data obtained with a particular product and therefore must be assigned on a case-by-case basis. A stated potency value can be determined from the mean potency estimates of several batches. This is the nominal potency value of the batches, and can be regarded as the expected potency for all batches. Acceptance limits must be established around this value for batch release (see Fig. 3). Vials or ampoules of product can be labelled with the stated potency as long as the mean potency estimates lie within the acceptance limits and the assay performance criteria are fulfilled. It is usually inappropriate to label vials of a biological product made using biotechnology with individual batch mean potency estimates.

Conclusions

Bioassays are often criticized as being «difficult», variable, tedious, expensive, time consuming and non-specific. However, they alone directly reflect and quantify the biological activity of a biological substance. Although other methods, e.g.

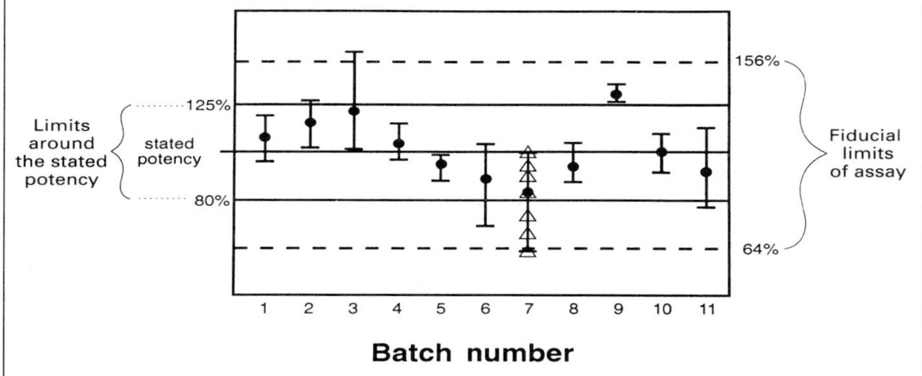

Fig. 3: An example of how statistical limits are applied to biological assays. A biological should be assayed in several independent bioassays. Each independent assay will provide an estimate of the potency of a sample from a specified batch of the biological (△). A mean potency (●) for the batch is obtained from the independent estimates for that batch and fiducial limits about that mean are obtained using the variability among the log potency estimates. The limits around the stated potency are determined on the basis of required accuracy of biological administered to patients. The fiducial limits are set to allow for the variation inherent in bioassays and to ensure confidence in the calculated mean value for any batch.

immunochemical or physicochemical procedures can be «easier» to carry out, they do not provide the same type of information as bioassays (see Table 3). They are therefore not«alternatives» to bioassays, but complementary methods.

A relatively common misconception is that bioassays are so variable and imprecise that the results obtained are not usable for quantitative purposes. Although this might apply if a bad choice of assay is made, it can be avoided by careful selection of bioassay method, format and analysis. Carefully designed, validated and correctly analysed bioassays can provide quantitative information which is as good as (or even better than) that produced using other non-biological assay procedures. Development of new bioassay systems such as those using receptor-transfected cell lines [8] or based on direct measurement of biochemical intermediates involved in signal transduction processes [9] may improve the characteristics of bioassays. Thus, bioassays provide crucial data at all stages of product development. Bioassay data are essential as part of product characterization and complement information obtained using other procedures.

To consider a biotechnology-produced biological product as «well-characterized», a characterization package is usually required which includes data derived from a variety of different complementary procedures. This should include appropriate bioassay derived data, particularly for product potency, stability and batch-to-batch consistency.

Table 3: Type of information provided by different assay systems for biological products produced by biotechnology.

Procedure	Data provided/Inferred						
	Size	Charge	1° Structure	2°/3° Structure	4° Structure	Purity	Potency
HPLC:							
Size exclusion	+++	–	–	++	+++	+++	–
Ion exchange	–	++++++	+++	–	+	+++	–
Hydrophobic interaction	+	+	+++	++	++	+++	–
Reverse phase HPLC	+	+	++++	++	–	++++	–
Circular dichroism	–	–	–	++++	+++	++	–
NMR	+	–	++++	++++++	–	++	–
Mass Spectrometry	++++++	+	++++++	–	–	++++	–
SDS gel electrophoresis:							
Reducing conditions	+++	–	+++	–	–	+++	–
Non-reducing conditions	–	–	++	+++	+++	+++	–
Isoelectric focussing	–	++++	+++	+	+	+++	–
Immunoblotting	+++	+	++	+++	+++	–	–
Immunoassays	–	–	+/–	+/–	+/–	–	–
Receptor binding	–	–	++	+++	+++	–	++
Bioassays	–	–	+++	++++	++++	–/+	++++++

Acknowledgements

We thank Helen Tivnann for data, Alan Heath and Rose Gaines-Das for advice on statistics, and Deborah Richards for processing the manuscript.

References

1 Mire-Sluis A, Gerrard T, Gaines Das R, Padilla A, Thorpe R: Biological Assays: Their role in the development and quality control of recombinant biological medicinal products. Biologicals 1996;24:351-362.

2 Thorpe R, Wadhwa M, Bird CR, Mire-Sluis AR: Detection and Measurement of Cytokines. Blood Rev 1992;6:133-148.

3 Bristow A, Gaines-Das R, Jeffcoate S, Schulster D: The first international standard for somatropin: Report of an international collaborative study. Growth Regulation 1995;5:133-141.

4 Wadhwa M, Bird C, Page LA, Mire-Sluis AR, Thorpe R: Quantitative biological assays for individual cytokines, in Balkwill FR (ed): Cytokines – A Practical Approach, 2nd Edition. Oxford, IRL Press, 1995, pp 357-391.

5 Mire-Sluis AR, Page L, Thorpe R: Quantitative cell line based bioassays for human cytokines. J Immunol Methods 1995;187:191-199.

6 Mire-Sluis AR, Gaines-Das R, Thorpe R, Participants in the study: The International standard for Granulocyte Colony Stimulating Factor (G-CSF): Evaluation in a collaborative study. J Immunol Methods 1995;179:117-126.

7 Finney DJ: Statistical methods in Biological Assay (3rd Edition). London, Charles Griffin & Co Ltd, 1978.

8 Hu-Li J, Ohara J, Watson C, Tsang W, Paul WE: Derivation of a T cell line that is highly responsive to IL-4 and IL-2 (CT.4R) and of an IL-2 hyporesponsive mutant of that line (CT.4S). J Immunol 1989;142:800-807.

9 Sadick M, Sliwkowski M, Nuijens A, Bald L, Chiang N, Lofgren J, Wong W: Analysis of heregulin-induced ErbB2 phosphorylation with a high-throughput kinase receptor activation enzyme-linked immunosorbant assay. Anal Biochem 1996;235:207-214.

Dr. R. Thorpe, Division of Immunology, N.I.B.S.C., Blanche Lane, South Mimms, Potters Bar, Herts, EN6 3QG, U.K.

Brown F, Fernandez J (eds): Development of Specifications for Biotechnology Pharmaceutical Products. Dev Biol Stand. Basel, Karger, 1997, vol 91, p 89

WORKSHOP III

In-Process Testing/Limits

Brown F, Fernandez J (eds): Development of Specifications for Biotechnology Pharmaceutical Products. Dev Biol Stand. Basel, Karger, 1997, vol 91, pp 91-93

WORKSHOP SUMMARY

J. Fernandez

The workshop on «In-Process Testing Limits» targeted the evolution of the process as the source of limits that may be applied. Information derived from process development, process validation and statistical process control are used in conjunction with assay development and batch data to derive tentative limits for the process and its product intermediates. By the time the product reaches licensure all specifications and limits must be well defined and justified. In cases where data do not follow a normal distribution, statistical process control may not be applicable. It is important to maintain tracking of product yield and take action should the yield vary outside the norms of production.

Process monitoring should focus on safety of the product with regard to bioburden. An aseptic process needs differentiation from a low bioburden product. Endotoxin can be introduced during processing and steps need to be taken to ensure endotoxin limits are controlled on buffers and other product contact materials. Risk analyses of the process should be performed as early as possible following the commitment to a final process sequence. During the risk analyses of the process sequence, hold times for various steps need to be justified and supported with stability data on the intermediates. Consolidating multiple process streams into a single product pool requires data to support the varying hold times of the different process streams. Analyses of critical steps relative to the integrity of the process intermediate may require extensive data collection at the process step. Raw materials may also require characterization of their stability profiles in addition to routine qualifying and release. All raw materials need to pass specifications at material expiry; however, chemical stability can be shown for aged raw materials by demonstrating once that it can be used successfully to make an intermediate or bulk. Excipient raw materials require enough study to ensure their stability profiles at their expiration date.

Workshop Slides

In-Process Testing/Limits – Slide 1

- In-process specifications should be based on information derived from process development, process validation and statistical process control.
- As assays are developed and batch information is collected, tentative limits can be considered. At licensure specifications and limits are finalized.
- Statistical process control may not apply in cases where a normal distribution doesn't exist. Common sense supercedes.

In-Process Testing/Limits – Slide 2

- Product yield is a meaningful parameter for Biotechnology products. Action should be taken when trends are outside the norm.
- No need to trend everything. Appropriate parameters should be measured (i.e. specific productivity gives different information than titer).

In-Process Testing/Limits – Slide 3

- The primary consideration should be safety. With regard to bioburden, acceptable levels are a function of the step in the process and what the product is being used for. An aseptic process product needs to be differentiated from a low bioburden product.
- Endotoxin can be introduced at any time during processing. Therefore, limits should be set from the beginning especially for buffers.
- A risk analysis should be performed as *early* as possible.

In-Process Testing/Limits – Slide 4

- What are allowable product hold times? At what point is stability data required to support hold steps?
- If hold times are brief, no stability studies are necessary. However, if a purification scheme indicates pooling or multiple eluents are being held at a single step, data should support the storage time required.

In-Process Testing/Limits – Slide 5

- If a process intermediate is removed at a critical step, a test should be performed there rather than at the final vial step. If adequate data has been collected, which indicates acceptable removal, a test need only be performed at re-validation.

In-Process Testing/Limits – Slide 6

- In addition to qualifying and releasing a given raw material for use, the stability profile of the raw material may be quite important, particularly with final product excipients.
- Every raw material needs written specifications which should be met at expiry, however, this is dependent on usage. Chemical stability can be shown for an aged raw material by demonstrating once that it can be successfully used to make an intermediate or final bulk. More studies may be required for excipient raw materials.

Brown F, Fernandez J (eds): Development of Specifications for Biotechnology Pharmaceutical Products. Dev Biol Stand. Basel, Karger, 1997, vol 91, p 95

WORKSHOP IV

Reference Standards/Materials

Brown F, Fernandez J (eds): Development of Specifications for Biotechnology Pharmaceutical Products. Dev Biol Stand. Basel, Karger, 1997, vol 91, pp 97-101

Workshop Summary

J. Fernandez

The workshop on «Reference Standards/Materials» opened with questioning the purpose of a reference standard or material. When determining the need for a reference material or standard, these questions need to be asked:

(i) What is the purpose of this reference material?

(ii) What quantity is needed?

(iii) How would we obtain the material?

(iv) Who is the end user?

(v) When is the material needed?

Having answered those questions, an appropriate candidate can be selected and characterized. Defining whether or not such a reference material becomes a «gold standard» vs. a working standard becomes an exercise in the quality of the characterization as well as the degree of characterization. Standards may be differentiated according to purpose; i.e. chemical assay, bioassay, impurity assay, or a material used in determining system suitability.

The use of «units» or IU of a standard is being replaced by use of the mass of a standard. Real time stability is preferred over calculated/theoretical stability. Reference standards should not carry an expiration date, but should define the optimal storage conditions for the user. As standards evolve in their characterization, the continuity between the existing standard and its replacement by a more characterized material needs to be maintained. Establishing a protocol for defining the character of the standard and its appropriate statistical calculations is critical. The range of standards can include Primary standards, Regional standards, and National standards.

In defining the reference materials used with regard to «well characterized proteins» various materials are used at various stages of product characterization. An «In-House» reference material can be used to define an acceptable impurity profile for the protein and may be formulated different from the final product to allow maintenance of the material integrity. An «In-House» reference material may also be used in evaluation of release assays and further characterization of the final product. For products in development, an early qualification lot may be a suitable reference material. Use of the lot release data as a baseline for future product evaluation allows development of a database for the reference material being used in the product characterization. For marketed products, a typical production lot may be used as an internal reference vs. use of a reference standard supplied by national/international reference laboratories. Characterization of the internal standard vs. the national/international reference material is imperative if the national/international standard is not used routinely.

Characterization assays depend on the intended use of the reference material and should be as exhaustive as possible to provide for extensive information on the molecule. Characterization should include correlation to the protein structure wherever possible. Sources of international standards include WHO (NIBSC) and the Pharmacopoeias. A major issue in establishing International Standards is the extensive time that it takes to establish a standard. Issues in establishing International Standards include the source of the standard (natural vs. recombinant,) and potency vs. bio-identity. The regulatory agencies feel that common working standards need to be established for well characterized products. In the case of monoclonal antibodies, regulatory agencies indicate that establishment of international reference materials may not be necessary.

The Pharmacopoeias have established Reference Standards Programmes where the need for a standard is determined by committee when a monograph outlines an analytical procedure that requires the establishment of a reference standard. The USP establishes standards at 100% and uses a collaborative process to characterize the standards. USP reference standards have no expiry dates. International Standards are used to define biotechnology product potencies per unit of activity. Candidate standards are obtained from as many sources as possible and a collaborative study system is used to define the standards. Quality markers are built into the studies to ensure integrity of the qualification process.

In summary, establishment of standards throughout the product development process is critical to maintaining the ability to qualify the product. Evolution of internal standards occurs as the process and product become well defined and, as the product becomes marketed, the reference standards become a vital part of maintaining product quality.

Workshop Slides

Reference Standards/Materials – Slide 1

- When, how, why, for what, how many and for whom?
- Select an appropriate candidate
- Characterization
- A «Gold Standard» vs a working standard
- Types of standards: assay, bioassays, impurity, system suitability

Reference Standards/Materials – Slide 2

- Expression of content is moving from I.U. to mass
- Preference is for real-time stability vs theoretical
- No expiry date
- Optimal storage conditions
- Evolution of standard should consider continuity
- The establishment of your standard should include protocol and be statistically correct
- The types of standards: primary standards, regional standards, and national standards

Reference Standards/Materials – Slide 3

- Reference materials with regard to «Well Characterized Proteins»
- In-house reference material:
- – Acceptable impurity profile
- – During evaluation one must perform release assays and characterization
- – Formulation may be different from product
- – Use for all testing

Reference Standards/Materials – Slide 4

- Development products:
- – Use one of the early lots (Qual Lot)
- – Use lot release data

- Marketed products:
- – Use a combination of lots (bulk) or a typical lot
- – A potency standard vs. Biochem standards
- – Perform lot release testing (some tests multiple times) and characterization

Reference Standards/Materials – Slide 5

- Characterization assays depend on:
- – intended use
- – as exhaustive as possible
- – correlation to structure

Reference Standards/Materials – Slide 6

- International standards:
- – WHO (NIBSC) Pharmacopoeias

- One issue with International Standards is the length of time to establish

Reference Standards/Materials – Slide 7

- Source of standard may be natural or biotech which may result in differences
- Potency vs Bio-identity
- Need to establish common working standards for well characterized products. Agencies feel that this is needed.

Reference Standards/Materials – Slide 8

- The need for international standards for monoclonal antibody products – Agencies indicated that this may not be necessary

- Pharmacopoeias reference standards program:
- – Determined as a result by request by committee
- – A monograph will drive the need for a standard because of an analytical procedure that needs a reference

Reference Standards/Materials – Slide 9

- – USP standards are established at 100%
- – A collaborative study is performed to characterize the standard
- – Has no expiry date

- International standards:
- – For potency tests only to define a unit of activity
- – Candidate from as many sources as possible
- – Collaborative study with multiple labs

- Quality markers built into the study

Reference Standards/Materials – Slide 10

- Standards:
- – Good practice for use
- – Need for both Bioassay (IUs) standards and strength standards
- – Case by case situation

Brown F, Fernandez J (eds): Development of Specifications for Biotechnology Pharmaceutical Products. Dev Biol Stand. Basel, Karger, 1997, vol 91, p 103

SESSION IV

Role of the Pharmacopoeia and Validation in Specification Development

Moderator: *Jean Huxsoll, M.S.*

Brown F, Fernandez J (eds): Development of Specifications for Biotechnology Pharmaceutical Products. Dev Biol Stand. Basel, Karger, 1997, vol 91, pp 105-113

The Role of Assay Validation in Specification Development

R.A. Baffi

Genentech, Inc., South San Francisco, CA, USA

Key words: Assay validation, method validation, specifications, validation parameters.

Abstract: Three of the main components that contribute to the establishment of specifications relate to manufacturing consistency, clinical experience, and assay performance. The validation of an analytical control system for biotechnology pharmaceutical products represents a diverse analytical challenge. This paper focusses on the key components of assay validation and how they influence specification design. In particular, issues relating to precision, accuracy, and specificity are discussed in the context of their influence and impact on specification design. Specific examples are presented that encompass assays that are used for purity and potency and their interrelationships.

INTRODUCTION

The development and validation of an analytical test method are in and of themselves unrelated to specifications. It is only when one evaluates the interrelationship between the process that is used to make the product and the biological relevance of the impurities detected by the specific method that one can begin to appreciate the impact that assay validation has on the design of specifications. For example, one may develop and validate an exquisite method for the detection and quantitation of a specific product impurity. However, if the process that is used to make the product is validated to consistently and reproducibly remove that impurity, then the utility of that specific method as part of the specification and control system needs to be challenged. In addition, the information gained in clinical trials can impact the types of methods and, ultimately, the specifications that are necessary. For example, the presence of a product-related impurity may be allowable at relatively high levels even if this causes reduced specific activity as long as an appropriate dose response has been demonstrated in clinical trials and the product is shown to be efficacious.

This paper, therefore, focusses less on how to validate specific methods and more on how that validation influences the design of specifications. The discussion begins with the various types of methods that may be used followed by a discussion of the critical elements to be evaluated during assay validation. Lastly, there is a discussion on the interrelationships of three assays used in the analytical control system for Nutropin AQ®, the first liquid formulation of recombinant human growth hormone.

Types of Methods

The development and validation of an analytical method share some similarities with the artwork of Rube Goldberg. Goldberg's drawings of absurdly connected machines that actually seemed to work became so well known that Webster's Dictionary added «Rube Goldberg» to its listing, defining it as «accomplishing by complex means what seemingly could be done simply». Although most machines work to make difficult tasks simple, it is clear how difficult Goldberg designed «closing a window if it starts to rain while you are away from home» [1]. This cartoon is particularly relevant because Goldberg anticipated the emergence of biotechnology in his design. One may see the similarities between the complexity of Rube's machine and the design of a peptide map, in which the simple question is asked, «What is the identity of a protein?»

In a very Rube Goldbergian manner, we propose a solution in which we first: reconstitute, dilute, reduce, carboxymethylate, re-reduce, buffer exchange, digest, quench, chromatograph, and visually and mathematically compare the elution profile of the sample and co-mix to the elution profile of the reference material. This comparison covers peak retention times, number of peaks, and relative peak response while demonstrating the absence of any new significant peaks. After this is successfully accomplished, we can state «Positive Identity».

With all the complexity inherent in the development and validation of a peptide map, it is still capable of performing its intended task (Fig. 1). A single amino acid

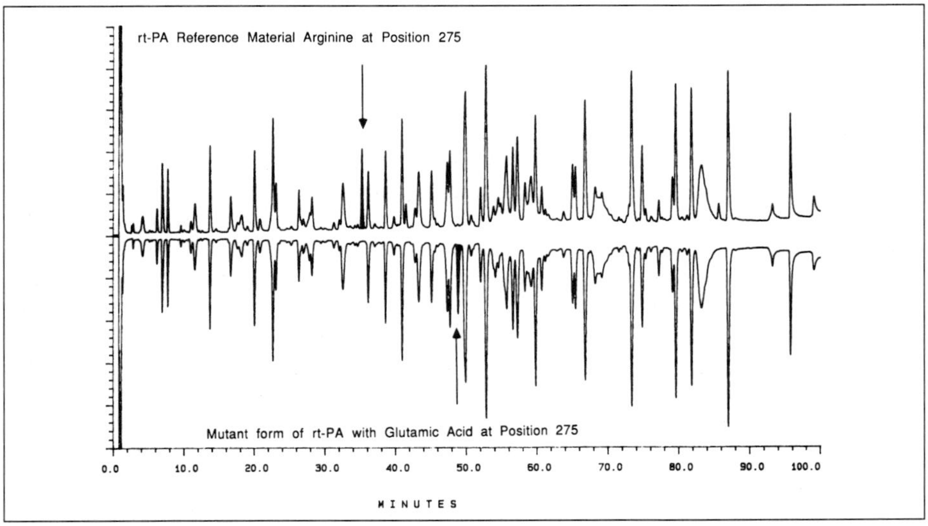

Fig. 1: Tryptic maps of recombinant tissue plasminogen activator (rt-PA) reference material (top) and a mutant form of rt-PA (bottom), which has a glutamic acid residue in place of the normal arginine residue at position 275. Arrows illustrate the differences in the two chromatograms caused by the amino acid substitution. (Reprinted with permission of Marcel Dekker, Inc.)

substitution in a variant of recombinant tissue plasminogen activator (rt-PA) was detected by this method [2]. Replacement of arginine by glutamic acid results in the detection of a new peptide peak not present in the reference material and a concurrent loss in the arginine-containing peptide peak. It is not possible in this paper to discuss further the analytical challenges in the validation of peptide maps. However, a Pharmaceutical Research and Manufacturers of America (PhRMA) «White Paper» devoted to this topic was submitted recently for publication [3] in fulfillment of an International Conference on Harmonisation (ICH) initiative on Genetic Stability.

There are several different types of methods that can be employed to establish product specifications. These methods may be used at different stages of manufacturing, such as in-process, bulk drug substance, and/or final drug product. These methods include ones that are used for identification, detection and quantitation of impurities (process or product related), quantitation of the product, microbiological methods for the assessment of adventitious agents (i.e., virus and mycoplasma), and potency assays for the determination of specific activity. This paper discusses in some detail assays specifically related to impurities and potency. At least one assay from each category is described below and it will be obvious that it is not unusual for a single assay to serve two purposes, i.e., it may be used to assess both the purity and quantity of a product.

Analytical Method Validation Criteria

Analytical method validation criteria are the elements that, to some extent, dictate what type of specification may be considered for a product. They include the following: specificity, detection and quantitation limits, precision, accuracy, range, linearity, robustness, and stability-indicating properties, as described in the ICH documents on assay validation [4, 5]. This paper focusses on accuracy, precision, and specificity. This does not mean that other criteria are not important nor is it meant to underestimate the role those other criteria play in assay validation and specification design.

Accuracy and precision are two criteria that are often confused. Accuracy expresses the closeness of agreement between the value that is accepted (either as a conventional true value or as an accepted reference value) and the value found. Precision expresses the closeness of agreement among a series of measurements obtained from multiple sampling of the same homogeneous sample under prescribed conditions. Therefore, one can describe four types of assays that may be developed. The most desirable is the accurate and precise method, which may be described as a cluster of analytical results very close to the intended target value. These types of assays lend themselves to the establishment of quantitative specifications. Examples of these are chromatographic- or spectrophotometric-based assays. Another category of assays is the accurate but imprecise assay, in which the analytical results are very close to the intended target but the variation around the target is large. These types of assays lend themselves to the establishment of limit-based specifications. Examples of these are animal bioassays and immunoassays. Another category of assays is one that is inaccurate but very precise. An assay of this type may make it difficult to establish specifications unless the inaccuracy is extremely reproducible and therefore correctable by some predetermined factor, or the magnitude of the inaccuracy is relatively small. The least desirable assays are ones that are inaccurate and imprecise. In fact, there may be no value in this type of assay because it would be difficult, if not impossible, to validate it.

Specificity is a critical attribute of an assay. Specificity addresses the ability of an assay to assess the analyte in the presence of components that may be expected to be present (see Fig. 2). These components may be impurities, degradation compounds, or excipients that are likely to be found in the product. The ability of methods to resolve product-related impurities adequately is crucial for the evaluation of the overall purity level of the product. One may have to rely on these methods to resolve product-related impurities that are only slightly different from the intended product. These methods may be relied upon to purify adequate quantities of these product-related impurities to assess biological activity as related to the intended product.

Specific Examples

This discussion will now focus on some specific examples of assay validations and their impact on the rationale for specification design, specifically some methods used to monitor the quality of Nutropin AQ®. As previously mentioned, this was the first liquid formulated human growth hormone product. The formulation was designed with a specific goal in mind, which was to improve patient compliance by designing a product that was easier to use when administered compared with lyophilized forms of human growth hormone. This was a formidable task in light of the propensity for human growth hormone to aggregate and deamidate. The goal was met by evaluating the consequence of various impurities and degradation products from a safety and efficacy perspective and then designing the process, formulation, and system to control those of greatest concern adequately.

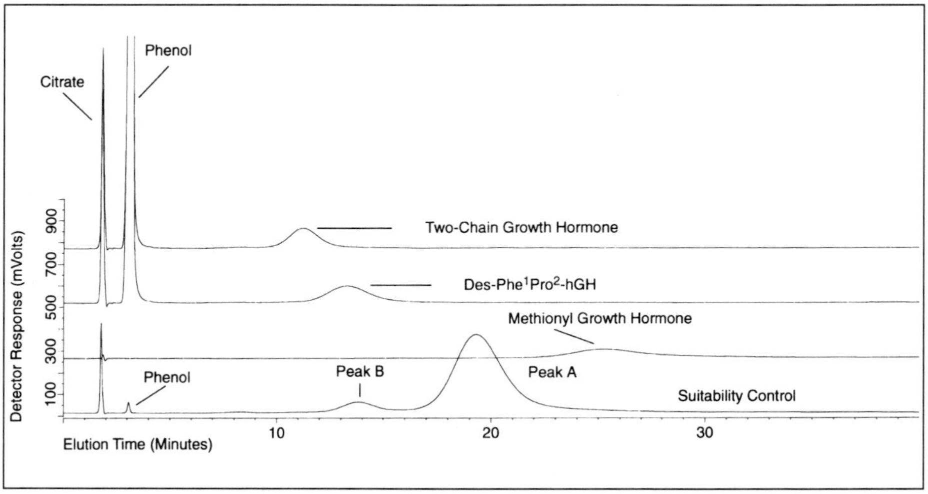

Fig. 2: A chromatogram generated by RPHPLC showing the specificity of the method in detecting various forms of human growth hormone.

The first method for consideration is high-performance size-exclusion chromatography (HPSEC). This methodology separates proteins based on molecular size or, more specifically, hydrodynamic volume (see Fig. 3). Within the analytical control system for Nutropin AQ®, this method serves two functions. First, it is used to determine the purity of the product because it can distinguish growth hormone monomer from dimers and higher molecular weight aggregates. The method was shown to have adequate specificity in that it can clearly resolve monomeric, dimeric, and aggregated forms of growth hormone and can quantitate them. In addition, it is used to assess protein concentration because Nutropin AQ® is a multiuse product and is preserved with phenol. This method allows for the separation of the phenol from the protein and thereby allows for quantitation of protein concentration.

Another valuable method is high-performance ion-exchange chromatography (HPIEC) (see Fig. 4). This methodology separates proteins based on charge. Within the analytical control system for Nutropin AQ®, this method is used to determine the purity of the product because it can distinguish deamidated forms of growth hormone and the intermediates that lead to deamidation. The method was shown to have adequate specificity in that it can clearly resolve mono- and di-deamidated forms of growth hormone and the succinimide intermediate from growth hormone itself. It has more than adequate sensitivity to quantitatively detect these forms of growth hormone (see Fig. 5).

The ability to quantify aggregated and deamidated forms of growth hormone adequately is an academic exercise up to the point at which one places some relevance on the importance of these product-related impurities of growth hormone. The relevance of these forms of growth hormone becomes clear when the specific activity of each of them is evaluated. The specific activity of each of these forms of growth hormone was evaluated by three different assays.

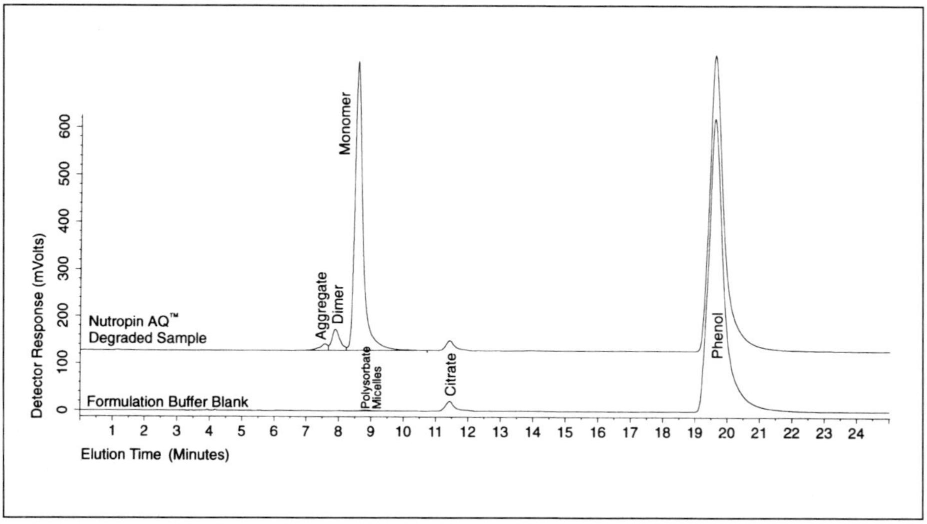

Fig. 3: A chromatogram generated by HPSEC showing the specificity of the method in resolving high molecular weight aggregate, dimer, and monomer of human growth hormone.

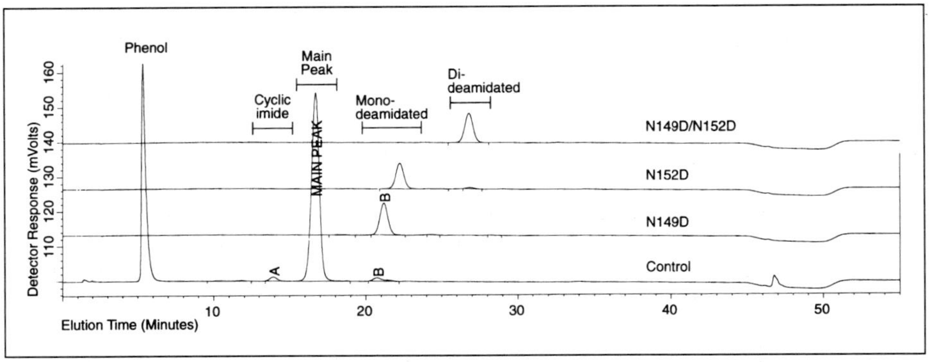

Fig. 4: A chromatogram generated by HPIEC showing the specificity of the method in detecting growth hormone charge variants. Shown are representative chromatograms of a growth hormone control lot and the elution profiles of three aspartate mutants (Asn to Asp), N149D, N152D, and N149D/N152D, which depict peaks analogous to the growth hormone main peak, a cyclic imide, a mono-deamidated, and a di-deamidated growth hormone species.

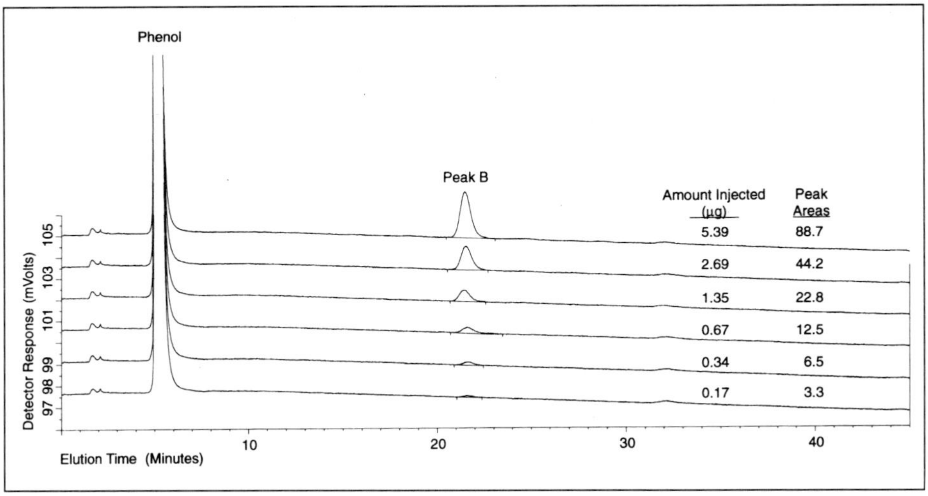

Fig. 5: Limit of detection of HPIEC. Shown are the elution profiles obtained from injections of a serially diluted sample of a mono-deamidated growth hormone variant over the range of 0.17 µg to 5.39 µg per 20-µL injection.

The first assay is the animal-based rat weight gain bioassay, which suffers from poor precision although it is highly accurate [6]. The second assay is the cell-based cell proliferation assay, and it is not only very accurate, but has improved precision when compared to the rat weight gain bioassay. In addition, it has the advantage of using cells that will only respond to human growth hormone, and thus it also has greater specificity. The third assay is chromatographic, the high-performance receptor binding chromatography (HPRBC) assay, and it is the most accurate and precise of the three. This method does have some limitations; it only monitors the binding of growth hormone to its receptor without demonstrating any biological function. Nonetheless, the information gained from HPRBC analysis is very informative because it correlates very well with the other two methods as described below.

The accuracy and precision of the rat weight gain bioassay are illustrated in Figure 6, which graphically presents results obtained for a single lot of human growth hormone assayed 23 times. The mean value obtained for relative potency (specific activity) was extremely close to the theoretical value of 1.0. Relative potency is expressed as the variation of the results at two standard deviations about the mean. Therefore, assessment of the 95% confidence interval ranges from a low of 0.7 to a high of 1.4. This precision primarily relegates the use of this assay to that of a limit test; that is, based on the high number of determinations required to obtain the true value, the resultant specification is established to ensure that the relative potency of a given sample exceeds an established minimum. As shown in Table 1, results obtained for the cell proliferation assay and the HPRBC assay are as accurate as the rat weight gain bioassay but much more precise. Nonetheless, trends in the rat weight gain bioassay can be correlated with assays of greater precision.

Table 1: Growth hormone potency values obtained from different assays.

	Rat bioassay	Receptor binding assay	Cell proliferation assay
Mean (IU/vial)	13.0	12.9	13.2
RSD (%)	23.0	4.1	9.0
Range (IU/vial)	9–20	11.7–13.6	12.0–14.8

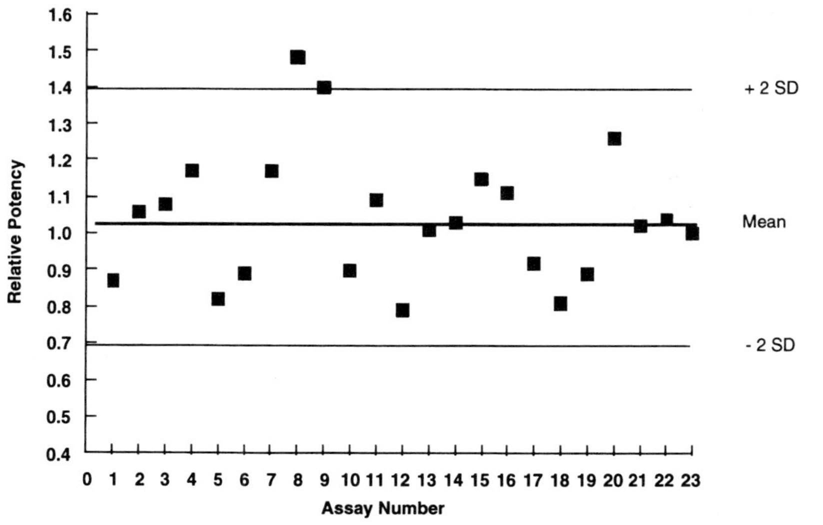

Fig. 6: Relative potency from the analysis of one lot of human growth hormone in the rat weight gain bioassay. The sample was analysed 23 times.

Several degraded or variant forms of growth hormone were assayed by all three methods (see Table 2). The results are very interesting because they relate to the HPSEC and the HPIEC methods previously described. The relative potency of the covalent dimer is very low in all three assays, whereas the values for the deamidated version of growth hormone are virtually indistinguishable from reference material. The analytical procedures for the detection of dimeric and deamidated forms of growth hormone have similar accuracy and precision because of their chromatographic similarities. However, from a biological and efficacy perspective, the relevance of each of these variants of growth hormone is significantly different. Therefore, it is not surprising that specifications for dimeric and deamidated forms of growth hormone are based more on clinical significance and manufacturing consistency, respectively, than on assay precision. In fact, the specification for dimeric forms of growth hormone is, appropriately, much tighter than that established for the deamidated forms.

Table 2: Bioactivity of genetic and chemical variants of growth hormone as measured by different assays.

Sample	**Rat bioassay**	**Receptor binding assay**	**Cell proliferation assay**
Deamidated rhGH	0.88	1.01	0.86
Aspartic acid rhGH (N149D)	0.80	1.00	0.87
Aspartic acid rhGH (N152D)	1.22	0.95	0.94
Aspartic acid rhGH (N149D+N152D)	0.85	0.96	0.93
Met-hGH covalent dimer	0.10	0.07	0.36

In conclusion, this paper highlights the methods that may be used in an analytical control system and the elements necessary for assay validation. Specific examples of methodology were discussed, where the overall accuracy, precision, and other elements of validation were shown to be comparable, yet different specifications were established based on the clinical significance of the variant that was detected by each method. Finally, the interrelationships of assay validation, clinical experience, and manufacturing consistency were presented as used in the establishment of a portion of the analytical control system for Nutropin AQ®, the first liquid formulation of recombinant human growth hormone.

Glossary

Accuracy: Expresses the closeness of agreement between the value that is accepted either as a conventional true value or an accepted reference value and the value found.

Detection limit: The lowest amount of analyte in a sample that can be detected, but not necessarily quantitated, as an exact value.

Identification tests: Those tests that are intended to ensure the identity of an analyte in a sample. This is normally achieved by comparison of a property of the sample to that of a reference standard.

Impurity tests: Those tests that are intended to accurately reflect the purity characteristics of the sample. These tests can be either quantitative or limit based.

Linearity: The ability (within a given range) to obtain test results that are directly proportional to the concentration of analyte in the sample.

Microbiological tests: Those tests that demonstrate the presence, quantity, or character of microbes.

Potency/Bioassay/Binding tests: Those tests that demonstrate some biological attribute(s) of an analyte. These tests may include animal-, cell-, or biochemical-based determinations.

Precision: The closeness of agreement among a series of measurements obtained from multiple sampling of the same homogeneous sample under prescribed conditions.

Quantitation limit: The lowest amount of analyte in a sample that can be quantitatively determined with suitable precision and accuracy.

Quantitative tests: Those tests that are intended to accurately measure the amount of analyte in a sample.

Range: The interval between the upper and lower concentration of analyte in the sample for which it has been demonstrated that the analytical procedure has a suitable level of precision, accuracy, and linearity.

Robustness: The measure of an analytical procedure's capacity to remain unaffected by small, but deliberate, variations in method parameters and provides an indication of its reliability during normal usage.

Specificity: The ability to assess the analyte in the presence of components that may be expected to be present.

Stability-indicating properties: The ability of a test procedure to discern changes in an analyte over time at the recommended storage conditions and to discern changes in an analyte that has been subjected to stress conditions.

References

1 Keller C, compiler. The best of Rube Goldberg. Englewood Cliffs (NJ): Prentice-Hall, 1979, p 52.

2 Garnick RL, Ross MJ, du Mée CP: Analysis of recombinant biologicals, in Swarbrick J, Boylan JC (eds): Encyclopedia of Pharmaceutical Technology. New York, Marcel Dekker, 1988, vol 1, pp 253–313.

3 Biologics and Biotechnology Section, Pharmaceutical Research and Manufacturers of America (PhRMA). Validation of peptide mapping for protein identity and genetic stability. Biologicals 1996;24:255-275.

4 International Conference on Harmonisation. Validation of analytical procedures: methodology, step 2. The Third International Conference on Harmonisation of Technical Requirements for Registration of Pharmaceuticals for Human Use; 1995 Nov 29–Dec 1; Yokohama, Japan. Geneva: ICH Secretariat (c/o IFPMA), 1995.

5 International Conference on Harmonisation. Guideline on validation of analytical procedures: definitions and terminology. Fed Regist 1995 (Mar 1); 60(40):11260–11262.

6 Roswall EC, Mukku VR, Chen AB, Hoff EH, Chu H, McKay PA et al: Novel assays based on human growth hormone receptor as alternatives to the rat weight gain bioassay for recombinant human growth hormone. Biologicals 1996;24:25–39.

Dr. R.A. Baffi, Genentech, Inc., DNA Way, South San Francisco, CA 94080, USA.

List of Speakers

ANICETTI V.R., Genentech, Inc., MS 65, 460 Point San Bruno Blvd., San Francisco, CA 94080, USA.

BAFFI R.A, Genentech, Inc., MS 65, 460 Point San Bruno Blvd., San Francisco, CA 94080, USA.

DAVIS G.C,. Eli Lilly & Co., Lilly Corporate Center, Indianopolis, IN 46285, USA.

FERNANDEZ J.L., Berlex Biosciences, 15049 San Pablo Ave., Richmond, CA 94804-0099, USA.

GARNICK R., Genentech, Inc., MS 87, 460 Point San Bruno Blvd., South San Francisco, CA 94080, USA.

GEIGERT J., IDEC Pharmaceuticals Corp., 11011 Torreyana Rd., San Diego, CA 92121, USA.

HAYAKAWA T., National Institute of Health Sciences, 1-18-1, Kumiyoga, Setagaya-Ku, 158 Tokyo, Japan.

HORAUD F., Institut Pasteur, 28, rue du Dr. Roux, 75725 Paris, Cedex 15, France.

HUXSOLL J.F., Matrix Pharmaceutical, Inc., 34700 Campus Dr., Fremont, CA 94555, USA.

LUBINIECKI A.A., Smithkline Beecham Pharm., UE-0428, 709 Swedeland Rd., King of Prussia, PA 19406, USA.

MURANO G., CBER, 1401 Rockville Pike, Rockville, MD 20832, USA.

PATTERSON S., Chiron Corporation, 4560 Horton Street, Emeryville, CA 94608-2916, USA.

SJOHOLM I., Medical Products Agency, Husargatan 8, S-75103 Uppsala, Sweden.

THORPE R., NIBSC, Blanche Lane, South Mimms, Potters Bar, EN6 3QG Hertfordshire, U.K.

TROUVIN J.-H., Agence du Médicament, 143-147 Blvd Anatole France, F-93285 Saint-Denis, Cedex, France.

Index of Authors

INTERNATIONAL ASSOCIATION OF BIOLOGICAL STANDARDIZATION

WHY NOT BECOME A MEMBER OF THE IABS?

As a member of the IABS, it is possible to contribute directly to the activity of our Association and to support the research, development, standardization and regulation of biological products. By your membership you would:

- receive our quarterly *Newsletter* which gives information on the activities of the IABS and highlights international news of current interest;
- subscribe to our journal *Biologicals*, published quarterly by Academic Press (London), at a reduced rate;
- take advantage of the discount on booksellers' prices for the volumes in the symposium series *Developments in Biological Standardization*, published by Karger;
- participate fully in our international meetings, with a reduction in the registration fees.

If you are not already a member of our Association and wish to join us, please complete this form and send it to our permanent office:

BIOSTANDARDS
P.O.B. 456
78, av. de la Roseraie
CH-1211 Geneva 4
Switzerland
Fax: (41) 22 702 93 55

or hand it to the Secretariat, if attending a Meeting.

- -

Surname.. First name(s)..

Scientific titles... Professional position...

Complete mailing address..

....

Field of expertise...

Tel.. Fax...

Annual fee: Sfr. 180.– which includes membership, subscription to *Biologicals* and the quarterly *Newsletter* of the IABS;
or Sfr. 90.-, without the subscription to *Biologicals*.

Method of payment:

❍ I have given instructions to my bank to make a credit transfer to the Swiss Bank Corporation, P.O.B. CH-1211 Geneva 11, Switzerland, a/c Biostandards n° 131,652.0.
❍ I enclose my cheque for Sfr............................... made payable to Biostandards
❍ I authorise you to charge Sfr.................. to my card: VISA / MASTERCARD / AM. EXPRESS

N°.. Expiry date.........................

Signature...

Symposia Series in Immunobiological Standardization

Vol. 1: Talloires 1965	Rabies
Vol. 2: Munich 1965	Neurovirulence of viral vaccines
Vol. 3: Stockholm 1965	Biotechnical developments in bacterial vaccine production
Vol. 4: Royaumont 1965	Immunological methods of biological standardization
Vol. 5: London 1966	Laboratory animals
Vol. 6: Utrecht 1966	Adjuvants of immunity
Vol. 7: Marburg/L. 1967	Assay of combined antigens
Vol. 8: Lyon 1967	Foot-and-Mouth Disease: Variants and immunity
Vol. 9: Paris 1967	Pseudotuberculosis
Vol. 10: London 1967	Biological assay methods of vaccine and sera
22nd Symp., London 1968	Standardization of immunofluorescence. Ed. Blackwell
Vol. 11: London 1968	Rubella vaccines
Vol. 12: Tunis 1968	Brucellosis. Standardization and control of vaccines and reagents
Vol. 13: Utrecht 1969	Pertussis vaccine
Vol. 14: London 1969	Standardization of interferon and interferon inducers
Vol. 15: Berne 1969	Enterobacterial vaccines
Vol. 16: Versailles 1970	Anti-lymphocytic serum
29th Symp., Brighton 1969	Hazards of handling simians. Ed. Lab. Animals Ltd. (4)
Vol. 17: Frankfurt 1970	BCG vaccine
Vol. 18: Copenhagen 1972	HL-A reagents
Vol. 19: Utrecht 1972	Smallpox vaccines
Vol. 20: London 1972	Influenza
Vol. 21: Lyon 1972	Rabies II
Vol. 22: Monaco 1973	The use of vaccines

Developments in Biological Standardization

Vol. 23: Madrid 1973	Sterilization and sterility testing of biological substances
Vol. 24: San Francisco 1973	Preservatives in biological products
Vol. 25: Lyon 1973	Requirements for poultry virus vaccines
Vol. 26: Budapest 1973	Selected veterinary vaccines
Vol. 27: Budapest 1973	Purity of human plasma proteins
Vol. 28: London 1974	Immunity to infections of the respiratory system in man and animals
46th Symp., London 1974	Breeding of simians and their uses in developmental biology. Ed. Lab. Animals Ltd. (6)
Vol. 29: Geneva 1974 (WHO/IABS)	Standardization and control of allergens administered to man
Vol. 30: Milan 1974	Viral hepatitis
Vol. 31: Rabat 1975	Brucellosis (II)
Vol. 32: Paris 1975 (OIE/IABS)	Colstridial sera and vaccines for veterinary use
Vol. 33: Douglas 1975	Vaccination by the non-parenteral route
Vol. 34: Budapest 1976	Pyrogenicity, innocuity and toxicity test systems for biological products
Vol. 35: Lyon 1976	Foot-and-mouth disease vaccines

Vol. 36: Lyon Washington D.C. 1976	Freeze-drying of biological products
Vol. 37: Geneva 1976	Standardization of cell substrates for the production of virus vaccines
Vol. 38: London 1977	Biological preparations in the treatment of cancer
Vol. 39: Geneva 1977	Influenza immunization (II)
Vol. 40: Marburg 1977	Standardization of rabies vaccines (III) (WHO/IABS)
Vol. 41: Guadeloupe 1978	Vaccinations in the developing countries
Vol. 42: Paris 1978	Second General Meeting of European Society of Animal Cell Technology
Vol. 43: Brussels 1978	Immunization: Benefit versus risk factors
Vol. 44: Villars 1979	Test methods for the quality control of plasma proteins
Vol. 45: San Antonio 1979	The Standardization of animals to improve biomedical research, production and control
Vol. 46: Oxford 1979	Third General Meeting of European Society of Animal Cell Technology
Vol. 47: Bilthoven 1980	Reassessment of inactivated poliomyelitis vaccine
Vol. 48: Geneva 1980	Standardization of albumin, plasma substitutes and plasmapheresis
Vol. 49: Leetown 1981	Fish biologics: Serodiagnostics and vaccines
Vol. 50: Heidelberg 1981	Joint ESACT-IABS Meeting on the use of heteroploid and other cell substrates for the production of biologicals
Vol. 51: Lyon 1981	Immunization of adult birds with inactivated oil adjuvant vaccines
Vol. 52: Lyon 1981	Herpes virus of man and animal: Standardization of immunological procedures
Vol. 53: Dublin 1982	Enteric infections in man and animals: Standardization of immunological procedures
Vol. 54: Athens 1982	Viral hepatitis: Standardization in immunoprophylaxis of infections by hepatitis viruses
Vol. 55: Copenhagen 1982	Fifth General Meeting of European Society of Animal Cell Technology
Vol. 56: Algeria 1983	IIIrd International Symposium on Brucellosis
Vol. 57: Paris 1983	Monoclonal antibodies: Standardization of their characterization and use
Vol. 58: Budapest 1983	BCG vaccines and tuberculins
Vol. 59: Geneva 1983	Standardization and control of biologicals produced by recombinant DNA technology
Vol. 60: Gardone Riviera 1983	Production and exploitation of existing and new animal cell substrates
Vol. 61: Geneva 1984	Proceedings of the Fourth International Symposium on pertussis
Vol. 62: Stockholm 1985	Diagnostics and vaccines for parasitic diseases
Vol. 63: San Francisco 1985	Use and standardization of chemically defined antigens
Vol. 64: London 1985	Reduction of animal usage in the development and control of biological products
Vol. 65: Amsterdam 1985	Use and standardization of combined vaccines
Vol. 66: Baden 1985	Advances in animal cell technology: cell engineering, evaluation and exploitation
Vol. 67: Melbourne 1986	Standardization in blood fractionation including coagulation factors
Vol. 68: Geneva 1986	Cells – Products – Safety
Vol. 69: London 1987	Cytokines: Laboratory and clinical evaluation
Vol. 70: Arlington, 1988	Continuous cell lines as substrates for biologicals
Vol. 71: Utrecht, 1989	Monoclonal antibodies for therapy, prevention and *in vivo* diagnosis of human disease
Vol. 72: Annecy, 1989	Progress in animal retroviruses
Vol. 73: Shizuoka, 1990	Pertussis: Evaluation and research on acellular pertussis vaccines
Vol. 74: Bethesda, MD, 1990	Biological product freeze-drying and formulation
Vol. 75: London, 1990	Virological aspects of the safety of biological products
Vol. 76: Bethesda, MD, 1991	Continuous cell lines – Current issues
Vol. 77: Annecy, 1991	Standardization of the immunopharmacology of natural and synthetic immunomodulators
Vol. 78: Bethesda, MD, 1991	Poliovirus attenuation: Molecular mechanisms and practical aspects
Vol. 79: Ploufragan, 1992	The first steps towards an international harmonization of veterinary biologicals: 1993 and free circulation of vaccines within the E.E.C.
Vol. 80: Heidelberg, 1992	Transmissible spongiform encephalopathies-Impact on animal and human health.
Vol. 81: Cannes, 1992	Virological safety aspects of plasma derivatives.
Vol. 82: Albany, NY, 1993	Recombinant vectors in vaccine development.
Vol. 83: Annecy, 1993	Genetic stability and recombinant product consistency.

Vol. 84: Langen, 1993	Non-target effects of live vaccines.
Vol. 85: Santa Fee, NM, 1994	Genetics of streptococci, enterococci and lactococci.
Vol. 86: Langen, 1994	Replacement, reduction and refinement of animal experiments in the development and control of biological products.
Vol. 87: Geneva, 1995	New Approaches to Stabilisation of Vaccines Potency.
Vol. 88: Bethesda, MD, 1995	Viral Safety and Evaluation of Viral Clearance from Biopharmaceutical Products.
Vol. 89: Rome, 1995	Pertussis Vaccine Trials.
Vol. 90: Oslo, 1996	Fish Vaccinology.

IMPRIME
RIE MEDE
CINE m+h
HYGIENE
octobre – 1997